THE
LOST EMPIRE
OF ATLANTIS

By the same author:

1421: The Year China Discovered the World

1434: The Year a Magnificent Chinese Fleet Sailed to Italy and Ignited the Renaissance

THE
LOST EMPIRE
OF ATLANTIS

||||||||||||||||||||||||||||||||||||||||||

HISTORY'S GREATEST
MYSTERY REVEALED

GAVIN MENZIES

WILLIAM MORROW
An Imprint of HarperCollins*Publishers*

If you would like to contact Gavin Menzies and the research team, please email them at zhenghe@gavinmenzies.net. Our website www.gavinmenzies.net is a focal point for ongoing research into pre-Columbian voyages to the New World. Please get in touch and join us on this great adventure!

First published in Great Britain in 2011 by Swordfish, an imprint of Orion Publishing Group, Ltd.

HarperCollins books may be purchased for educational, business, or sales promotional use. For information please write: Special Markets Department, HarperCollins Publishers, 10 East 53rd Street, New York, NY 10022.

FIRST U.S. EDITION

Library of Congress Cataloging-in-Publication Data has been applied for.

ISBN 978-0-06-204948-3

11 12 13 14 15 RRD 10 9 8 7 6 5 4 3 2 1

This book is dedicated to my beloved wife Marcella,
who has travelled with me on the journeys
related in this book and through life.

Here is her name in Minoan Linear A script.

CONTENTS

BOOK III: JOURNEYS WEST

BOOK IV: EXAMINING THE HEAVENS

BOOK V: THE REACHES OF EMPIRE

BOOK VI: THE LEGACY

ACKNOWLEDGEMENTS

Encouragement to write this book

This book, like *1421* and *1434*, is a collective endeavour. Hundreds of people, by and large friends via my websites, have encouraged me by persuading me that multiple intercontinental voyages were undertaken thousands of years before Columbus: indeed, long before Admiral Zheng He's voyages. So I should start by thanking those people who have taken the trouble to send me emails.

All of my books rely heavily on my experiences as a submarine navigator or captain. I am indebted to the Royal Navy for investing in me, training me for over a decade to fulfil those duties. I am particularly grateful to Admiral Sir John Woodward G.B.E., K.C.B., who trained me to be a submarine captain and who taught me to think laterally – that is, to address problems by examining the evidence rather than by using preconceived notions.

There have been many authors, far more distinguished and knowledgeable than I, whose books have been an inspiration to me. In their book *Pre-Columbian Contact with the Americas Across the Oceans: An Annotated Bibliography*, Emeritus Professor John L. Sorenson and Martin Raish have produced a summary of over 5,000 books or articles describing transcontinental voyages across the oceans over the past 8,000 years. Emeritus Professor Carl L. Johannessen, in a series of books and articles, has published similar accounts of intercontinental voyages over many millennia. Sorenson and Johannessen have joined forces to publish *World Trade and Biological Exchanges Before 1492*, which I have used time and again to provide evidence to support my claims in this book. Recently Professor Sorenson has published 'A Complex of Ritual and Ideology Shared by Mesoamerica and The Ancient Near East', a paper in which he sets out descriptions of thousands of intercontinental sea voyages thousands of years before Columbus.

Emeritus Professor John Coghlan has provided intellectual backing these five years, in support of a non-scholar 'who marches to the beat of a different drum'. John has had to face virulent criticism for having done so and I am most grateful for his unwavering backing.

There are authors whose view of history differs from that of established historians – I thank Professors Octave Du Temple and Roy Drier, Emeritus Professor James Scherz, James L. Guthrie and David Hoffman for their work on the ancient copper mines of Lake Superior and the missing millions of pounds of copper from those mines, which apparently vanished into thin air.

My story is about the Minoan fleets that travelled the oceans of the world before the ghastly explosion on Thera in 1450 BC, which wiped out the Minoan civilisation. Professor Spyridon Marinatos alerted us all to this adventure in 1964 when he chose to excavate the town of Akrotiri on Thera (Santorini), which had been a major Minoan base in the 2nd millennium BC. By good luck and judgement he stumbled upon the house of an admiral, buried in 1450 BC but still with intact walls. This gave the world its first sight of the superb Minoan ships which had then plied the world.

Professor Marinatos' excavations mirrored those of Sir Arthur Evans in Crete, where Sir Arthur is a legend. Single-handedly, through decades of excavations and research, he revealed the fabulous Minoan civilisation which burst upon the world in 3000 BC. I have relied heavily on Sir Arthur's work, not least in the context of the inheritance of classical Greece and the debt Greece and Europe owe to the Minoan civilisation. Sir Arthur's mantle has fallen upon Professor Stylianos Alexiou, whose book launched our adventure, and latterly Dr Minas Tsikritsis, whose work is described later in the book.

Many authors have spent a lifetime describing intercontinental voyages across the oceans in the third and second millennia BC. I would particularly like to thank Dr Gunnar Thompson for his accounts of the trade between America, Egypt and India, notably the trade in maize; Charlotte Rees and Liu Gang for their works on trade between America and China in the second and first millennia BC; David Hoffman for his researches into prehistoric voyages for copper between Europe and America, especially between the Great Lakes and the Atlantic; Tim Severin for demonstrating to us all that such voyages were possible; J. Lesley Fitton for knowledge of the trade between the Atlantic, Europe and the eastern Mediterranean; Professor Emeritus Bernard Knapp for his writings about the trade between Crete, Africa and the Levant; Dr Joan Aruz for mounting the superb exhibition 'Beyond Babylon' in the Metropolitan Museum of Art, New York (I have extensively referred to the beautiful book she edited relating to that exhibition); Professor Emeritus Manfred Bietak for his work on Minoan fleets in the Nile Delta; Professor Rao for unearthing the Bronze Age Indian port of Lothal; Professor Edward Keall of the Royal Ontario Museum for his team's excavation of the Bronze Age hoard in the Yemen; Hans Peter Duerr for his articles on Minoan trade with the Baltic in the 2nd millennium BC; Professor Beatriz Comendador Rey for her research and her team's excavations of Bronze Age seaborne voyages to Spain; Tony Hammond for his information about Bronze Age mining and trade in Britain; and Philip Coppens for his studies of the Great Lakes copper trade in the 2nd millennium BC.

Minoan shipbuilding expertise, which led to the intercontinental voyages without which there would have been no Atlantis civilisation, is at the heart of my story. Together with the rest of the world, I am indebted to Mr Mehmet Cakir who found the Uluburun wreck (c.1310 BC) and Professor Cemal Pulak who organised a very skilful series of dives over eleven summers, which have resulted in such a haul of evidence being taken from the seabed to the castle that has been adapted to house these amazing treasures. Professor Andreas Hauptmann and colleagues have analysed the chemical composition of the copper ingots in the Uluburun wreck and a number of other experts have carried out research into the goods, flora and fauna found in the wreck, as evidence of the voyage of the ship – notably Baltic amber; African ivory; shells from the Indian Ocean; and beads from India. Further thanks have been placed on my website.

Not only have I relied on the revolutionary research of those mentioned but I have equally depended upon the team without whom this book would never have been written. As in the past, Ian Hudson has co-ordinated the team with great skill and humour, integrating design work and Ms Moy's typing. Ms Moy of QED Secretarial Services has typed twenty-nine drafts speedily, accurately, economically and with good humour.

I owe a special tribute to Cedric Bell, who has supported my research in many ways for years. Originally a marine engineer, he has spent a lifetime in engineering. His roles have included those of surveyor, foundry engineer, works engineer and then production manager of Europe's largest lube oil plant. Following retirement he has spent fifteen years researching the Roman occupation of Britain on a full-time basis, finding many similarities between Chinese and Roman engineering. His contribution to this book has been enormous. Back in 2003 Cedric read *1421* shortly before visiting New Zealand. Several surveys then followed. These surveys proved that the Chinese had been mining and refining iron in New Zealand for 2,000 years. The evidence included harbours, wrecks, settlements and foundries. This led to a furore, followed by vitriolic attacks on Cedric by New Zealand 'historians'. I appointed a team of independent surveyors to check Cedric's finds by using ground-penetrating radar, sonics and the independent carbon dating of iron mortar and wood. The results are on my website. They show that Cedric's research was incredibly accurate.

As works maintenance engineer for Delta Metals in Birmingham, Cedric was responsible for a large non-ferrous foundry and extruders and an ore reclamation plant with ball mills, Wilfley tables and vacuum extraction flotation tanks. At the time, Delta produced 65 per cent of Britain's non-ferrous metals. Whenever I have come across a problem (there have been many) Cedric has either been able to answer me immediately or refer me to an expert who could.

He has also provided me with a stream of books, including the classic works on Bronze Age mining and smelting. Without his expert unfailing support, this book would not have been completed.

Luigi Bonomi, my literary agent, who has acted for me for the past ten years and has skilfully sold *1421* and *1434*, has been an inspiration. Luigi persuaded me to postpone my book dealing with Chinese voyages to the Americas in the 2nd millennium BC in favour of this one. Luigi has superb judgement on which I have relied throughout. Budding authors should beat a path to his door!

Luigi sold the world literary rights to this book to Orion – part of the Hachette Group, the world's largest publisher. Orion has been enormously supportive and enthusiastic. I should particularly like to thank Rowland White my publisher, his assistant Nicola Crossley and Susan Howe, foreign rights director and her team, as well as Helen Ewing and Georgie Widdrington.

Gaynor Aaltonen played a key role. She has skilfully turned my stilted prose into a readable book whilst at the same time incorporating an endless flood of new evidence, which has flowed on to our computers since Orion first took this book on. Without Gaynor's work there would not have been a book. I am indebted to her.

Finally Marcella, for without her unfailing kindness and support there would have been no *Lost Empire of Atlantis*. I and the book owe her everything. I am so happy the Minoan Linear A spelling of her name starts with the face of a cat!

Gavin Menzies
London
St Valentines Day 2011

LIST OF ILLUSTRATIONS
AND DIAGRAMS

The illustrations are inspired by the wonderful frescoes at Thera, Knossos and Tell el-Dab'a and drawn by Catherine Grant (www.catherinezoraida.com)

Book I: 'Minoan ladies in all their finery'
Book II: 'Lions pouncing on their prey at Tell el-Dab'a'
Book III: 'Minoan bull leaping – Knossos'
Book IV: 'The Prince of Lilies'
Book V: 'An ocean-going Minoan ship'
Book VI: 'The Phaestos Disc'

The following images from the frescoes are found throughout the book:

Swallow
Blue lion
Small boat from the 'flotilla' fresco
Fisherman
Antelope
Jumping deer
Partridges

LIST OF PLATES

First plate section

The Pyramid of Khufu, photograph by Digr

Syrian with his son and man of Keftiu with rhyton. Tomb of Menkheperraseneb, Egypt, Thebes, reign of Thutmose III. The Metropolitan Museum of Art, Rogers Fund, 1930 (30.4.55) © The Metropolitan Museum of Art

View of Palace of Knossos, photograph by Eigene Aufnahme

Arthur Evans' bust, photograph by Peterak

Bull head rhyton from Knossos, photograph by Jerzy Strzelecki

Phaestos disc side A, photograph by PRA

Phaestos disc Side B, photograph by PRA

View of Palace of Phaestos, photograph by Eigene Aufnahme

Axe in the form of a panther, Heraklion Museum, Crete © Nick Kaye www.flickr.com/people/nickkaye. All rights reserved

Dolphin fresco at Knossos © Getty Images

The throne room at Knossos, photograph by Lapplaender

Pithoi in storeroom at Knossos © 2000 Grisel Gonzalez. All Rights Reserved

Jewellery in the 'Aigina Treasure', the Master of Animals at the British Museum, photograph by Bkwillwm

Minoan bee brooch, Heraklion Museum, photograph by Andree Stephan

Golden bee pendant, Heraklion Museum, Crete © Nick Kaye

The Minoan flotilla, Ch. Doumas, The Wall Paintings of Thera, Idryma Theras–Petros M. Nomikos, Athens 1992

Uluburun wreck copper ingots and other artefacts, Bodrum Museum of Underwater Archaeology, photographs by Gavin Menzies

Al-Midamman bronze hoard photographs, Brian Boyle © Royal Ontario Museum

Second plate section

The Hagia Triada sarcophagus © Nick Kaye

Sculpted stone sunflower juxtaposed with live sunflower, Halebid, Karnataka, India © Carl L. Johannessen

Wall sculpture from Hoysala Dynasty Halebid temple at Somnathpur, India, showing maize ears © Carl L. Johannessen

Stone carving at Pattadakal temple, India, shows a parrot perched on a sunflower © Carl L. Johannessen

Stone carving of a pineapple in a cave temple in Udaiguri, India © The American Institute of Indian Studies

Francisco José de Goya y Lucientes, The Agility and Audacity of Juanito Apiñani in the Ring at Madrid, plate 20 from the series La Tauromaquia, 1814–1816. Etching and aquatint. Meadows Museum, SMU, Dallas, Algur H. Meadows Collection, MM.67.07.20. Photography by Michael Bodycomb

Minoan bull leaper, British Museum, photograph by Mike Peel

Minoan bull leaper, Heraklion Museum, photograph by Jerzy Strzelecki.

Dover ship © Dover Museum and Bronze Age Boat Gallery

The Nebra disc, photograph by Rainer Zenz

Stonehenge, photograph by Stefan Kühn

A selection of photographs from the Wiltshire Heritage Museum, Devizes includes:

    227: Upton Lovell G2 – Amber space-plate necklace with complex borings

    599: Wilton, Bronze looped palstave

    616: Rushall Down, Disc-headed bronze pin

    340: Upton Lovell G1, Faience beads

    159: Wilsford G56, Bronze dagger

    266: Winterbourne Stoke G5, Bronze dagger

    623: Found between Salisbury and Amesbury, Bronze bracelet

    166: Wilsford G23, Bronze crutch-headed pin with hollow, open-ended head

    237: Shrewton G27, Stone battle axe

    All images © Wiltshire Heritage Museum, Devizes

Comparisons of Stonehenge, Uluburun and Great Lakes copper tools
and implements:
Coiled snake effigy (Uluburun wreck and Great Lakes)
Animal weights at the British museum, the Uluburun wreck and
the Great Lakes.
Conical points (Uluburun wreck and Great Lakes)
Triangulate spear head (Uluburun wreck and Great Lakes)
Gaff hooks (Uluburun wreck and Great Lakes)
Bronze knives (Uluburun wreck and Great Lakes)
To view photographs of Great Lakes copper tools and come to your
own conclusions, please visit
www.copperculture.zoomshare.com

The Amesbury archer, Salisbury Museum, photograph by Ian Hudson
Antikythera device, photograph by Marsyas

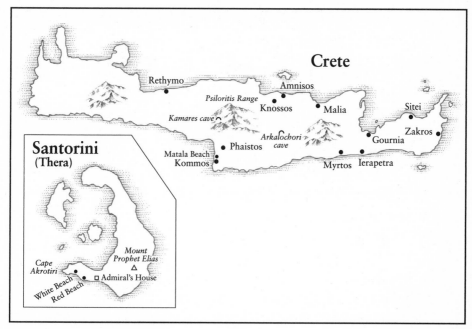

Minoan Crete and Santorini

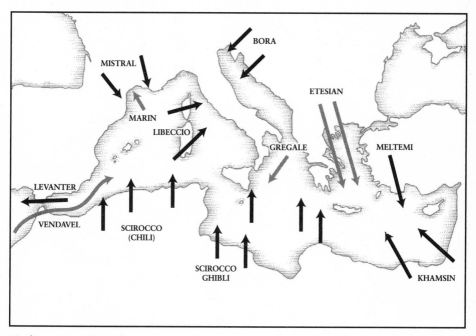

Mediterranean Winds

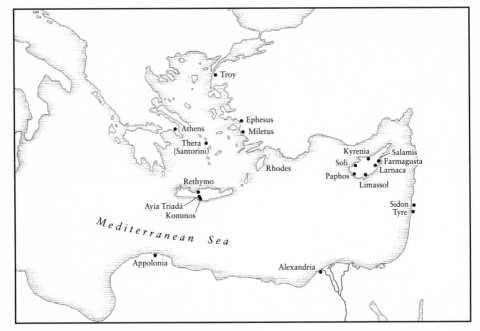

Minoan Trade Empire in the Mediterranean

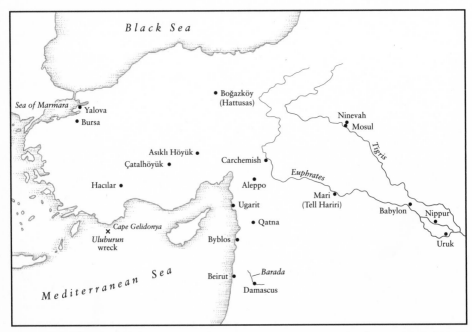

Turkey and the Near East

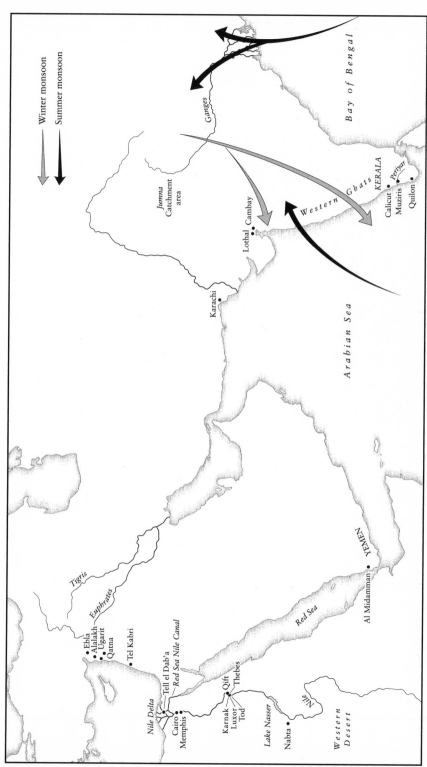

Egypt and route to India

Spain and Portugal

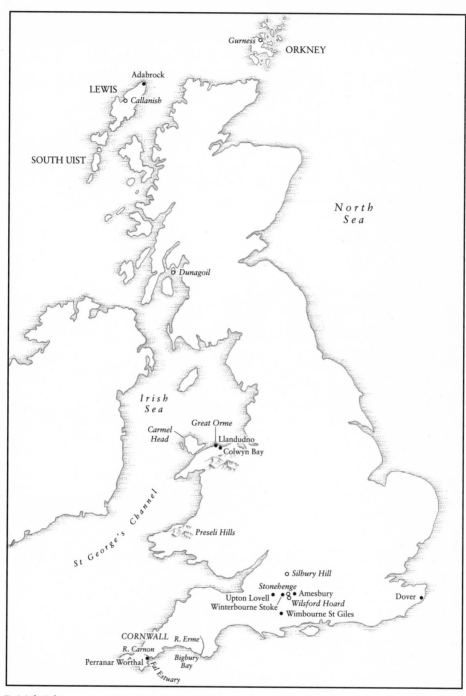

British Isles

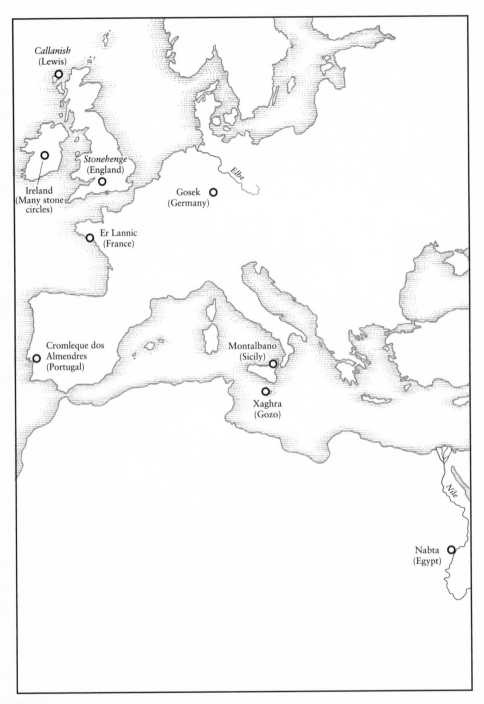

Stone circles found around the world in the Minoans' wake

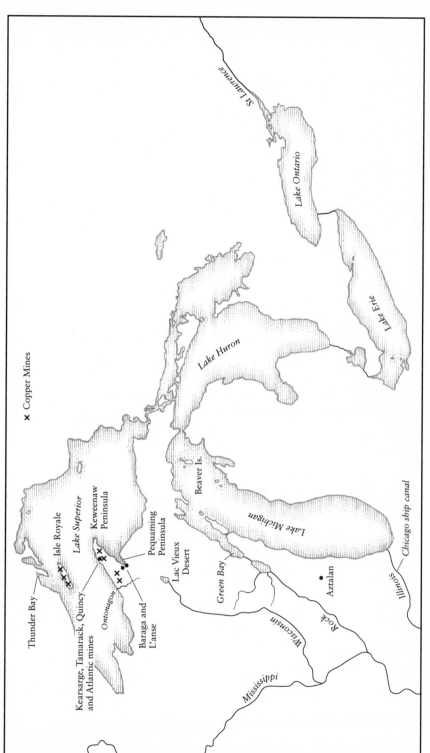

× Copper Mines

St Lawrence

Lake Ontario

Lake Erie

Lake Huron

Beaver Is.

Isle Royale

Lake Superior

Keweenaw
Peninsula

Pequaming
Peninsula

Thunder Bay

Kearsarge, Tamarack, Quincy
and Atlantic mines

Ontonagon

Baraga and
L'anse

Lac Vieux
Desert

Green Bay

Lake Michigan

Chicago ship canal

Illinois

Aztalan

Rock

Wisconsin

Mississippi

The Great Lakes

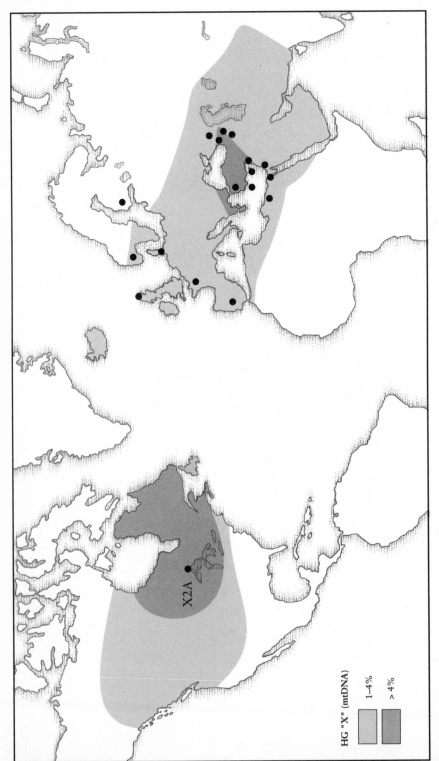

Distribution of Haplotype X2

HG "X" (mtDNA)

1-4%

> 4%

X2A

BOOK I

||||||||||||||||||||||||||||||

DISCOVERY

THE MINOAN
CIVILISATION

1

AN ADVENTURE ON CRETE

I was gazing north from the balcony of our hotel, the glowing lights of the town huddled at my feet. Far below, the Aegean stretched away through the night towards a lost horizon. Somewhere out there in the open sea lay the ancient, ruined island of Thera. What I did not know, as I turned and made my way back into the room, was that hidden beneath that island was a secret that was thousands of years old, a secret that would revolutionise my view of history. All of my ideas about history and world exploration were about to be turned upside down.

My wife Marcella and I had been looking forward to a quiet, contemplative Christmas, shut off from the world. After researching material for a new book I was dog-tired, so we decided to take a short break on the island of Crete. We would travel via Athens, one of our favourite places on earth. No more mobiles or email: we'd spend Christmas by candlelight in a former Byzantine monastery. Or so we thought. Our cosy dream of long walks through the classical ruins, followed by a week of Spartan simplicity in the mountains of Crete, was shattered when we arrived in Athens to find a mini riot in progress, right outside the hotel. In front of us protesting mobs of people milled around, shouting, waving placards and overturning cars. Police – bearing guns and wearing riot gear – stood menacingly in front of them.

So we took ourselves straight off to Crete, finding another spot, idyllic in its own way: a small hotel in the comfortable old Venetian port of Rethymno, in the north. Our first two days were marked by torrential downpours of rain. Then on Christmas Eve the rain lifted.

A few rays of sun turned into a soft mellow morning that beckoned us to explore beyond the snow-capped mountains we could see from our balcony. What we found that day put any thoughts of quiet, and of rest, out of the question: the discovery would set me on a determined hunt for knowledge around the globe.

We set out on a drive. In places, the narrow roads were more like fords, running with the past two days' rain. As we drove rather gingerly through a muddy series of pretty villages, we could see that grapes still hung on the vine. The roadsides were carpeted with yellow clover and the fig trees still had their deep green leaves, even at Christmas. Crete is an incredibly fertile island. As we left one rickety village, we saw two men drag a pig across the road and string it up between two ladders, ready for slaughter. After a journey along winding roads, often blocked by herds of black and white goats, their bells tinkling to ward off snakes, we pitched up at our destination, picked that morning at random: the ancient palace of Phaestos.

Myth has it that the city of Phaestos was founded by one of the sons of the legendary hero Hercules. It certainly looks majestic. The ruined palace complex unfolds like a great white plate on a deep green backcloth of pine forest to the south of the island. The ancient Greeks believed that this was one of the cities founded by the great King Minos, a mythological figure who reigned over Crete generations before the Trojan War. As we began to explore the ruins we quickly discovered that the symbol of Minos' royalty was the *labrys*, or double axe, a formidable ceremonial weapon shaped like a waxing and waning moon, set back to back. The other symbol of Minos' immense power was the terrifying image of a rampant bull.

We stepped out on to a ruin that was staggering. Phaestos is vast: bigger than the Holy Roman Emperor Charlemagne's royal palace at Aachen, and at least three times the size of London's Buckingham Palace. Its powerful but simple architecture is flawlessly constructed in elegant cut stone and it is laid out in what appears to be a harmonious plan. Wide, open staircases lead from the Theatre to the Bull Ring to the Royal Palace, and from there to smooth stone platforms that allow views of the ring of mountains beyond, the green plain sloping away to the distant sea. The overall effect was one of lightness and air. As the sun danced off the pools and courtyards, reflecting the azure sky, the whole site seemed like a mirage floating between heaven and earth. The serenity and scale of the place reminded us both instantly of the same thing: the monumental architecture of Egypt.

One other fact immediately gripped us. Like Knossos, its perhaps more famous sister complex further northeast of us, the palace at Phaestos is ancient, older by far than the magnificent Parthenon in Athens, built c.450 BC when classical Greece was at its height – the people of Phaestos had lived in luxury and comfort more than a millennium before that. The palace is as venerable as the Old Kingdom of the pharaohs of Egypt and as ancient as the great pyramids of Giza. The site had been inhabited, we discovered, since 4000 BC.

This came as a real surprise to me. How was it that I had heard so little of this extraordinary but relatively obscure palace, whose beauty rivalled that of India's Taj Mahal? What did I really know about the people who had built it: the people who are known as 'the Minoans'? As we walked across the baking hot palace court, we realised that while most Europeans were still living in primitive huts, the ancient Minoans were building palaces with paved streets, baths and functioning sewers. Unparalleled for its age, the Minoans' advanced engineering knowledge gave them a sophisticated lifestyle that put other contemporary 'civilisations' in the shade: they had intricate water-piping systems, water-tight drains, advanced air-flow management and even earthquake-resistant walls.

We climbed a huge ceremonial stair, the steps slightly slanted to allow rainwater to run off. At the end of a narrow corridor we suddenly found ourselves in an alabaster-lined room. Here, a light-well struck sunbeams off the silent walls. These rooms, once several storeys high, were known as the Queen's apartments, we were told. Phaestos' rulers enjoyed stunning marble walls in their palaces: the people lived healthy and refined lives in well-built stone houses. Secure granaries kept wheat and millet safe from rats and mice and reservoirs held water all year round. The inhabitants enjoyed warm baths and showers – the men and women bathed separately – while their toilets had running water. Cut stone, expertly placed, lined the aqueducts that brought hot and cold water from the natural warm and cold springs which surrounded the palace. Terracotta pipes, built in interlocking sections, provided a constant supply of water, probably pumped via a system of hydraulics. All in all, what we saw before us was a more advanced way of living than in contemporary Old Kingdom Egypt, Vedic India or Shang China.

We duly bought Professor Stylianos Alexiou's book, *Minoan Civilization*, for seven euros. As we looked through the text it became evident that in ancient times, as now, Crete was an island of magnetic attraction, the sort of place storytellers and poets would speak of with awe. This reverence had inspired potent legends, both about the place and the people who lived there. The head of the gods, Zeus, was said to have been born and died on Crete and it was here that another god, Dionysus, allegedly invented wine.  In fact a large number of the ancient Greek myths I had learned at school had actually originated on Crete – their power so great that the tales have lasted for millennia. Epic sagas like those spun by the Greek poet Homer in the 8th century BC had been told at family firesides for centuries before that. In Book 19 of his *Odyssey*, Homer writes reverently about Knossos as a fabulous city lost in legend. After reading a little bit more about the Minoan civilisation, I realised that he was absolutely right.

This remarkable civilisation did not confine itself to Crete. Like the swallows we could see flying over our heads, the Minoans were seasoned summer travellers. In fact, the guide told us, Old Kingdom Egyptian frescoes showing diplomatic envoys from the Keftiu – as the ancient Egyptians called the peoples from Crete – decorate the tombs of dignitaries from the time of Thutmose III, in the 18th Dynasty. They were carrying ritual vases for pouring oils. That meant that from around 1425 BC there would have been Minoan travellers in Egypt: a pretty staggering idea. I wondered idly if these intrepid ancients had inspired some of the age-old Greek mythical epics: tales of Jason sailing the seas with the Argonauts, or of Odysseus' full decade of seafaring and dodging dangers before he returned home to his loyal wife Penelope.

I began my own Odyssey more than twenty years ago. It started with the discovery of a little-known Venetian map. Zuane Pizzigano's carefully drawn 15th-century chart showed both Europe and the supposedly undiscovered islands of Puerto Rico and Guadeloupe. Medieval history had become a consuming interest for me and that chance find eventually led me to believe that the history of the world – specifically the history of mankind's navigation of its seas – would have to be radically rewritten. Pizzigano's map was drawn in 1421 – and that's what I called my first book.

I found that the Portuguese who had embarked on Europe's restless voyages of discovery, uncovering the partly cloaked face of the globe, were themselves relying on much older maps. This begged an obvious enough question: who drew them? The trail of evidence led me to the other side of the world, to a people we have long admired for their ingenuity and wisdom. There was only one nation at the time, I realised, which had the material resources and, most crucially, the ships to embark on an adventure of such ambition: China. The Chinese had circumnavigated the globe a century before Magellan, I argued. They had discovered America – and had reached Australia 350 years before Britain's Captain Cook. To a former sea captain, Pizzigano's extraordinary chart held a hidden message, like a codex waiting to be uncovered. Although it was pure luck that I had found it at all, that first clue led me all over Europe and then beyond, to Asia, on a true voyage of discovery.

⁜⁜⁜⁜⁜⁜⁜⁜⁜⁜⁜

Here I was again, this time in Crete, fascinated by a civilisation that appeared to have such depth, that must have been so important to the world, and yet was so little understood. In comparison to the impressive body of knowledge the world has accumulated about the lives of the ancient Egyptians, it seemed almost as if there was a giant conspiracy to keep the exotic, vibrant Minoan culture secret. On one level, I had already vaguely known about the remarkable archaeology of Crete. The brightness and brilliance of the culture that had produced it – of that I had simply no idea. The fact was that these fun-loving ancients – who lived so well, who went diving with porpoises and whose athletic young men leapt over bulls' backs – had also made some of the finest jewellery ever seen. And they had painted frescoes to rival the best the European Renaissance was to offer. All of this had completely escaped me. As if to rub in the irony of this yawning gap in my knowledge, I was not entirely new to Crete.

When I joined the Navy in the 1950s, Great Britain was still a world power. So in my early life as a sailor I had travelled with great fleets to sea bases all over the world – from the Americas to Australasia and China. In 1958, as a junior watch-keeping officer, I was on gun patrol around Crete's neighbour island of Cyprus; our mission was to prevent weapons being smuggled in by terrorists. The EOKA Greek Cypriot nationalists were fighting against British rule, and we were kept remorselessly busy. When HMS *Diamond* was given a week's leave on nearby Crete we anchored in Souda Bay and paid our respects to the fallen, at the magnificent Second World War memorial.

Crete, placed strategically between Europe, Africa and Asia, is a rectangular island that measures 155 miles (or 250 kilometres) from east to west and between 7 and 37 miles north to south. During the Second World War HMS *Orion*, a cruiser which my father later commanded, lost hundreds of men here to attacks from German fighter bombers. The island is a strategic prize and has been fought over for thousands of years. First of all the Mycenaeans, a people from what is now the Greek mainland, seem to have displaced the Minoans. Then the Greeks battled the Romans; the Byzantines fought the Venetians; the Venetians gave way to the Ottomans; and the Ottomans were ejected by the Cretans. Finally, the Germans and the Allies each fought desperately for Crete, both sides acting with exemplary courage and barbaric ferocity. Even today there is a NATO port and a joint Greek/US airbase at Souda Bay.

We marched up a disused railway track, which in spring was a carpet of glorious wild flowers. After the incessant rain of Cyprus the countryside made a vivid impression on us: the island's fertile plains are warmed by Mediterranean sunshine for nine months a year. That night fifty years ago a local farmer allowed us to camp in his field. He provided a lamb and a barrel of local Cretan wine; we built a fire and had a traditional sailor's sing-song. Cook Mifsud played the hurdy-gurdy; Chief Steward Vassalo recited poetry. Later, the farmer's charming daughter, Maria, took me on a walk. She wanted to show me a ruined stone city, which she said was close by. 'It is very old,' she told me. 'More than 2,000 years old – the other way in time.'

What she meant was that the site dated from 2,000 years *before Christ.*

That seemed extraordinary to me. I dimly remembered from my schoolboy studies that the heyday of ancient Greece had been some 1,500 years later than that, at around 500 BC. Much later on in life I worked out that Maria and I must have walked to Archanes, a village which may have been the summer retreat for those Minoans who lived sumptuously at Knossos. But as the next two months were a whirl of anchoring off remote islands, swimming with the local girls and absorbing as much of the local culture and the *dópio krasi,* or local Marisini wine, as we could find, I had soon forgotten about an abstract-sounding series of ancient dates.

Now here I was again, with our oak-coloured local guide insisting, as Maria had before him, that the civilisation behind his tiny island had been as important to the world as that of the Egyptians. What if they were both right?

It was a local businessman and amateur archaeologist, Minos Kalo-kairinos, a namesake of the fabled King Minos, who discovered the first and most famous of the ancient palaces of Crete – Knossos –

in 1878. Kalokairinos initially uncovered a large storeyard containing *pithoi* – huge, almost man-height vessels used to store olive oil. At a time when archaeology was in its infancy, Kalokairinos had brought up find after find, wonders that had been buried for centuries in the dark earth. Unfortunately for him, some local landowners stepped in and stopped his work. When the German archaeologist Heinrich Schliemann, the excavator of Troy and Mycenae, attempted to purchase the 'Kefala hill' he was put off by what he thought was an exorbitant price. Then in 1894 the pioneering British archaeologist Sir Arthur Evans heard about what was happening and applied for a licence, investing profits from his family's paper mill to win the right to excavate. The ruins Evans extracted from Crete's baked earth were those of a palace whose magnificence could scarcely be imagined.

Discoveries of other ancient palaces, towns and ports would follow that of Knossos over the course of the next century. Evans seemed to have uncovered an entire ancient civilisation; an entrancing race with an advanced and exotic culture. He dubbed them 'the Minoans'. Why had he opted for that name? Like others before him, Evans was seduced by the power of Greek myth.

Before 1900, when Evans began to uncover the spectacular palace, our only real knowledge of this ancient civilisation came from the extraordinary substratum of myth that surrounds the island, as well as the awed references to it by classical poets. By ancient Cretan tradition, the peak of Mount Juktas, which dominates the skyline looking south from Knossos, is said to hold the imprint of the upturned face of the mighty Zeus. It is as if he is buried there and supports the island with his slumbering body. Famously, the mythical home of the controlling ruler, King Minos, contained a vast underground labyrinth. The formidable Minos had been the patron of the great inventor, Daedalus – and, more alarmingly, according to legend, was the tyrant who exacted human tribute from the Athenian mainland. Within the labyrinth the King Minos of myth kept the terrifying Minotaur, a half-bull, half-man monster. Every year, Minos demanded youths from

Athens as tribute and imprisoned them in the labyrinth for the beast to feed upon.

As Evans was excavating at Knossos one of the workmen gave a terrified cry. The myth was surfacing from the dust. He had found a 'black devil', he shouted, shying away in horror from the object he had plucked out from the soil. In fact what he had uncovered was a remarkable, red-eyed bust of a bull's head – a sculpture of monumental power and menace, crowned by a huge set of horns. The sculpture was detailed, lifelike. It is said that when they pulled it from its ancient resting place, the bull's fierce eyes moved in their sockets. As they dug deeper, Evans and his team were amazed to discover that this beautiful hilltop palace did seem to have at its heart a genuine labyrinth – a maze of deep, underground tunnels – buried beneath it. Frescoes showing charging bulls added to the archaeologist's growing conviction: all of the evidence suggested that the people here had worshipped a bull god.

Thirty years earlier, Evans' fellow archaeologist Heinrich Schliemann had shocked the world when he had dramatically declared that he had 'gazed upon the face of Agamemnon'. He had been excavating at the citadel of Mycenae, while in search of the legendary heroes of the Trojan War. The claim that the German had found the actual body of the heroic figure of legend gave weight to a compelling idea – that the much loved ancient texts of Virgil's *Aeneid* and Homer's *Iliad*, if not literally true, had a strong basis in reality. A romantic, Arthur Evans was convinced that he had done the same thing for ancient Greek myth that Schliemann had done for Virgil and Homer's epics. He firmly believed that he had found the true home of the mythical King Minos and his evil Minotaur.

||||||||||||||||||||

Legend may have surrounded the island, but it was a very real people that Evans had plucked from the ancient shadows; a people who at their peak of prosperity around 2160 to 1500 BC had plainly commanded fabulous wealth and power. What was also intriguing was the society's high level of sophistication – almost modernity.

11

Men and women appear to have been equal. More than that, the Minoans seem to have worshipped a female goddess, as well as the bull.

Now here I was clutching a booklet that was making the astounding claim that Phaestos was as old as the oldest Egyptian pyramids: the palace was contemporary with the time of the Egyptian Old Kingdom (2686–2125 BC). It had been first built in the era of the pharaohs Khufu and Khafre and the Great Pyramid at Giza.

THE PALACES

Few civilisations have been so completely lost to history as that of the Minoans. This is partly because their remarkable palaces were destroyed not once, but twice. The principal palace of Knossos was first destroyed by fire around 1700 BC. The new palace built after that was more like an urban complex than a single palace, with parts of it five storeys high. It was not only a royal residence, but the heart of religious ceremony and political life on the island. It was also a manufacturing base for exports such as swords and pottery. It was a busy place: there were olive and wine presses and grain mills. The need for water was dealt with by an aqueduct that carried water from springs about 6 miles (10 kilometres) away, at Archanes.

The magnificent buildings and terraced gardens were grouped around a vast central courtyard, which was used for sacred nocturnal festivals, bull-jumping and torch-lit ecstatic dances, as the islanders worshipped their deities. The whole complex stretched out for 5 acres (2 hectares). An ingenious system of bays and light-wells brought cool, dappled light into the magnificent, colonnaded palace. They even had window panes, made of thin sheets of translucent alabaster. The buildings' columns were made from the trunks of cypress trees, which were painted red and then mounted on a plinth.

The double-headed axe, a symbol of the *minos*, or king, appears on many of its walls. The Greek term for the axe is a *labyros*, which gives the labyrinth its name. The palace, which contained 1,300 rooms, was so complex – such a mass of halls, corridors and chambers with vast underground storage chambers running beneath – that some speculate that the dark, oppressive labyrinth of fame and legend was in fact Knossos itself. The second destruction, which was originally thought to be by earthquake, happened around 1450 BC.

Life doesn't appear to have been too grim for the ordinary Minoan citizen. Heraklion Museum has a ceramic model that depicts delightful rows of cheerful houses in the town below, banded with bright colour. They played board games, such as a version of draughts, and barbecued food was prepared outdoors on charcoal braziers. In the countryside the better-off also had summer mansions.

The major palace complexes discovered on Crete to date are at Knossos, Phaestos, Malia and Kato Zakros. Zakros is a fifth of the size of Knossos. Most of the palaces seem to be oriented with the landscape. The immense age of the palaces poses a problem with dating them. One dating system is based on the architectural development of the palaces, dividing the Minoan period into the Prepalatial, Protopalatial, Neopalatial and Postpalatial periods.

There was more. Reading Professor Alexiou's book I discovered that the remarkable Egyptian sun king Akhenaten had owned many pieces of Minoan pottery and had installed them at his palace at Amarna. The implications of this idea halted me in my tracks. The professor, a highly distinguished figure in his field, was saying that the Minoans had not only travelled to ancient Egypt: they had even traded with the pharaohs. (See first colour plate section.)

> Fictional representations of Cretans, the famous Keftiu as the Egyptians called them, carrying zoomorphic rhyta (ritual vases for pouring) and other works of art typical of the Neo-Palatial period as gifts from Crete, decorate the tombs of dignitaries of the same dynasty. Finally fragments of Post-Palatial [Cretan] pottery [1400–1100 BC] found in the Palace of Amenophis IV or Akhenaten at Amarna (inhabited from 1375 BC) help to determine the beginning of the Post-Palatial era and also the date of the destruction of the palace [of Phaestos] since similar pottery was found on its floors.

Since starting research for my first book, *1421*, in 1988, nothing in those twenty years has surprised me more than Professor Alexiou's evidence of the long-standing maritime trade between Crete and Egypt from 1991 BC to 1400 BC. My own thesis had been that the Chinese had been the first navigators of the world, in the year 1421– AD!. But evidence of international trading many centuries and more *before* the birth of Christ put an entirely different perspective on my proposed book about Chinese voyages to the Americas. According to solidly researched and established archaeological evidence, the Minoans had travelled far beyond their native shores. What's more, said our guide proudly, the Minoans had achieved another significant 'first'. They had invented writing. This struck me as unlikely. Hadn't the Egyptians or the Sumerians got there before them?

When I challenged him, the guide pointed proudly to his handheld notes, covered with slightly crumpled pictures. On his papers you could see a picture of a strange red ceramic plate, covered in clear, white markings. It was an enigma, like nothing else I'd ever seen. The symbols didn't go right to left, or even left to right, like Chinese script. The path of this language, if that's what it was, went round and round, in a labyrinthine circle.

'It's a mystery: it's something we cannot understand.' Our guide traced the symbols with his finger.

'What is it called?' I asked, truly intrigued.

'The Phaestos Disc,' he replied. He explained that these impenetrable ancient letters or words could be man's first linear writing: perhaps they could be better described as the first use of printing, given the fact that the symbols were stamped into the clay surface before it was fired, around 1700 BC. Here then was a hidden language; a secret history written in ceramic. The disc could be the key to understanding an entire, lost civilisation. But it was totally unintelligible, he said, even to the experts. Looking closely, I could see that the circular plate had 241 symbols imprinted on its surface. Some of the pictograms looked just like sticks, or maybe they were a form of basic counting. Others were strange, involved and full of what looked like symbolic meaning; images of fish, fruit and even human heads.

The disc was discovered in a small basement room in 1903, near the depositories of the 'archive chamber' in the northeast apartments of the Phaestos palace. A few inches away from it lay a tablet known as PH-1, which carried the first discovery of a mysterious Cretan written language. It is now known as 'Linear A'. Like the disc, it has so far completely eluded translation, although the script has since been found on many other objects and artefacts at various sites in Crete. Linear A's first known use was here at Phaestos, and some experts believe that Linear A and the strange pictographic language of the disc are closely related. How extraordinary: to hold the key to a lost civilisation right there in your hands, but to be unable to make sense of it. To view the Phaestos disc, please go to the first colour plate section.

Another overriding thought was troubling me: if the Minoans were so advanced – as educated and artistic as the ancient Egyptians, a civilisation which had invented both writing and printing, as well as employing an astonishing realism in art many centuries before classical Greece – why was it that the world knew so little about them? The next, inevitable question was: what happened to the Minoans?

Our guide's dramatic response to that question was to start me on my new quest. Phaestos, its sister palace-city of Knossos and the other Minoan cities were all 'destroyed in a massive earthquake', he told us. It seems that this captivating society had abruptly disappeared from view around 1450 BC. Ever since Knossos re-emerged from the fertile soil of Crete people have wondered what on earth could have happened to extinguish the life force of those exotic cities of long ago, each with a palace culture so powerful that it inspired the enduring myths of the ancient Greeks. Now I too was fired up to find out more.

'Are you sure it was an earthquake?' we asked the guide. 'Because it says in this book that the island of Santorini, or ancient Thera, about 90 miles away from here, was destroyed at the same time, by a huge volcano.' We had to follow the trail.

2

UNDER THE VOLCANO

Sixty-nine miles north of Crete lay an island in the Aegean Sea that might, we thought, yield the answers to the sudden disappearance of the Minoans. We were off to Santorini: known in ancient times as Thera.

We arrived late at night, exhausted, delayed by strong northerly winds. The Med, usually so calm and peaceful, can catch you out: within six hours of a 30-knot wind, the waves can reach 3 metres (10 feet) high. The Meltemi in particular is a violent wind which whips up in summer with little warning, often creating havoc with ships and ferries alike (see map). After that journey, even a short taxi ride felt like an expedition. Tired as we were, Marcella and I were excited as we approached our new destination: the Apanemo Hotel, built high on a rocky promontory. Although it was dark and we could hardly see a thing, we were finally here, safe, on an island that the world's first historian, Herodotus, called 'Kalliste' or 'The Most Fair'.

We awoke the next morning to a marvellous view overlooking the central lagoon, the so-called caldera or 'cooking pot', the volcano that had once been Santorini's molten heart. It sits on top of the most active volcanic centre in what is known as the South Aegean Volcanic Arc and we could see the effects of its violent geological past from our colourful Greek-style bedroom. The great crater blown out from its middle is spectacular: about 7.5 by 4.3 miles (12 by 7 kilometres) long and surrounded by steep cliffs 300 metres (980 feet) high on three sides. The water in the centre of the lagoon is nearly 400 metres (1,300 feet) deep, making it an extraordinarily safe harbour for all kinds of shipping. The islets of the upstart volcanic islands, Nea Kameni and Palae Kameni, lay right in front of us, the choppy narrow channel between them churned up by the northerly winds that had plagued us the day before. Far below us, two white cruise liners half the size of matchsticks were entering the deep blue of the caldera.

Today, the island's shape resembles a giant round black fruit cake with the centre gouged out, leaving a circular rim of inky volcanic soil surrounding the central lagoon. Like the icing sugar on a wedding cake, white villages tumble down in terraces and then cling for dear life to the caldera's circular rim and dramatic cliffs.

By chance I knew Thera reasonably well but only, as it were, from underneath. In the1960s I was navigator of the submarine HMS *Narwhal*, attached for two months to the Greek navy, which had asked us to take periscope photographs in the caldera.

A submerged submarine is the same weight as that of the volume of water it displaces. The weight of this volume of water varies with temperature and salinity – the warmer the water, the lighter the submarine must become. To maintain neutral buoyancy the submarine's weight is altered by pumping out or flooding in water. Submarines are sensitive – 455 litres (100 gallons) of water was all that was required to achieve *Narwhal*'s correct weight when the submarine was moving slowly. (Though *Narwhal* displaced 3,000 tons when dived.)

There are to this day underground springs and volcanic fissures which spew hot water and magma into the base of Thera's lagoon. We did not know how these would affect the temperature and salinity of the lagoon and hence the amount of water we would need to pump out of the submarine as we travelled at periscope depth. Neither were we sure if we had enough pumping capacity to deal with the caldera.

The problem was accentuated because we would move over the hot volcanic springs as we navigated the caldera. That meant that the submarine's buoyancy could be continuously changing. The entrance channel to the lagoon was narrow – wide enough for us, but tight. The narrowest part of the deep entrance to the caldera was some 183 metres (600 feet) wide – about the length of a ferry sideways on. However, the channel here was nearly 305 metres (1,000 feet) deep – plenty of room to get underneath passing ferries, so we were reasonably relaxed.

After the reconnaissance, we set off back to Zakinthos for a rendezvous with some Greek girls on the beach at sunset. They were going to show us where turtles came ashore to mate and lay their eggs. Then we would dance in the moonlight. That wonderful evening, I have to say, was my most enduring memory of Thera. I'd forgotten about anything else for over forty years.

All of this was to change. Over a breakfast of honey and rich Greek yoghurt, the hotel owner told us about the extraordinary city that lies beneath modern-day Santorini. It was discovered through the determination of one man, the Greek archaeologist Professor Spyridon Marinatos. The professor already knew Crete very well; so well, in fact, that in the 1930s he had made one of the island's most significant finds there, at what is known as the Arkalochori cave. At that site, Marinatos found a priceless hoard of ancient bronze goods and weapons, as well as one of the most famous double axes ever discovered on Crete, the impressive bronze votive offering known as the Arkalochori axe. That intrigued me. Bronze weapons must have been highly valuable: why hoard them away?

For years, Marinatos had entertained a hunch that there would be an ancient town on Santorini, of a similar date to those on Crete. Marinatos was an inspired archaeologist, whose long and exciting career eventually included excavations at world-famous sites such as Marathon and Thermopylae. He also followed Schliemann and others by undertaking research at the Bronze Age city of Mycenae – but discovering Thera was arguably Marinatos' most inspired moment.

It was a chance observation that led to Marinatos' extraordinary discovery. Convention had it that it was an earthquake that had destroyed Crete's palaces and towns. Yet one day, when the professor was excavating a villa, he realised that the entire interior of the house was filled with volcanic pumice. No earthquake could have done that. He was digging at Amnisos, a Minoan port town due north of Knossos, which lies due south of ancient Thera and its volcano. Looking with new eyes, Marinatos could also see that blocks of stone had been dragged apart, as if by a huge body of water: the pumice, meanwhile, was mixed with beach sand, as if everything had been thrown up in the air together. The prevailing earthquake theory must, he thought, be flawed. Looking at the crumbling archaeological material in his hands, Marinatos saw a tale of violence and destruction, but not the sort caused by an earthquake. This kind of devastation, he thought, must have come from the sea.

It also seemed to him that the villa and the surrounding ancient Minoan town had both been destroyed in the course of a few minutes, suffering a sudden and catastrophic fate that was directly comparable with the disaster that had befallen the ancient city of Pompeii. The Roman city had been buried by Mount Vesuvius in AD 79, a volcano so powerful that when it erupts large areas of southern Europe are choked with suffocating layers of ash.

Aware that there had been an almighty volcanic eruption on Santorini/Thera c.1450 BC, the archaeologist's gut instinct told him that something significant – and something from the same era as Knossos – could be buried there. It took him decades, but finally he started a dig just outside the modern town of Akrotiri, where we were now staying. His patience was rewarded almost straight away when, helped by the advice of a local man who knew the fields, he struck archaeological gold.

Burrowing through many feet of volcanic pumice is a tricky job – and back-breaking work. But not for Marinatos, apparently. It was as if he were a man possessed. He chose the exact right spot, a place where the hardened layer of volcanic pumice thinned to about 5 metres. Like Schliemann at ancient Troy, his hunch had proven itself to be 100 per cent correct. He had discovered an abandoned city.

||||||||||||||||||||

After breakfast we sat on a sunny wall and read what the dig director, Professor Christos Doumas, had to say about the volcano and the island of Santorini. At one stage the island was called Strongili, 'The Round One', but volcanic eruptions have turned its former round, bun-like shape into a croissant. Much later in history, the conquering armies of ancient Sparta gave the crescent of islands the name Thera. According to Professor Doumas:

> . . . In the last 400,000 years there have been more than 100 eruptions on this island, each adding a new layer of earth and rock, slowly making the island bigger. The last of these truly catastrophic eruptions came 3,600 years ago. . . .

As we read on, we realised that there was a true miracle – and that Marinatos had found an entire lost city – right beneath our feet. It dated back to at least 1450 BC, the year it's thought that the volcano erupted. Logically, that ancient city, now buried under mounds of volcanic ash, must have existed at roughly the same time as the palace of Phaestos. This was astounding: two such urbane societies no more than 120 miles (195 kilometres) from one another and thriving during an age we had regarded as plain 'prehistoric'.

At this stage, we were strictly tourists – I had no plan to write about Crete, or for that matter Thera; our interest was the pure enjoyment of an ancient enigma. Sadly, today's crowds of eager tourists can't visit the ruins of the underground world that Marinatos saw. Less than a third of the archaeological site has been excavated and the site is still dangerous, given the risk of falls. But we knew from our hotel owner that some of the frescoes unearthed so far have been restored and are in the town museum. We sauntered in the next morning to see them. After that moment, all research for my current book would be abandoned, as fascination with this ancient puzzle grew and gripped me.

The frescoes that had been buried beneath the lava-drowned island of Thera are painted as if drawn from life. Brilliantly coloured, delicate and remarkably lifelike, they show a fertile island rich with plants and wildlife. There were people in this ancient painted paradise: beautiful people, who had lived a life of luxury. We had both assumed that 3,000 years ago life was a matter of pure survival, not of fun.

When you think of the Bronze Age you tend to imagine grunting 'prehistoric' men and women living in caves, wearing animal skins, clubbing one another and going generally unwashed. Yet here, at the very centre of the Mediterranean Sea, was a glittering and highly advanced society. There was nothing cave-man-like about it, even with the idea of bull worship. In fact it looked wonderful. The people in the frescoes were striking, to say the least. Women had tight uplifting bodices, slit to the waist to reveal their breasts, like those on the frescoes at Knossos. The men were athletic, long-limbed and handsome and both men and women wore jewellery – lots of it, from earrings and armbands to necklaces. As on Crete, living standards were precocious: the Therans had fountains, flush toilets and bathtubs. The red, white and black stone houses stacked neatly against the ancient hillsides seemed almost better built and certainly more finely decorated than our flimsy modern 'little boxes'.

Who were these ancient Therans, we wondered? And why did their frescoes remind us so much of the murals on Crete? Noticing a huge *pithos*, or storage jar, that looked just like those we'd seen at Phaestos, I began to realise that Thera must have had a strong connection to the Minoans. Thousands of shards of fine Cretan-style ceramics have been found here, of what experts call 'the highest palatial quality' – used, they think, in rituals.[1]

Was this civilisation, too, Minoan? Even to my untutored gaze, there were strong similarities in the culture, art and architecture of Thera and Crete. Just like the Minoans on Crete, the people of Thera had a taste for spectacle, for music, for festivals and for fun. There were both cultural and spiritual connections. Here, too, the bull was

Notes are at the end of Part 1 on page 70.

a creature of cult worship and their worship of female gods was as striking as that of the Minoans. The very word 'Crete', etymologists say, has links to today's Greek term, 'strong goddess'. It seems that at Knossos each king, or 'Minos', married the moon-priestess.[2] At Phaestos and Knossos, women were as athletic as men: in a famous painting of athletes somersaulting over a bull, two girls are holding the bull. Here, too, women were portrayed as important, possibly even ruling figures.

We moved on to the next series of frescoes. The prehistoric paintings on walls, floors and vases were stunning: flowing, filled with colour and life. What was more, the beautifully drawn,

lavishly coloured pictures were stylistically just like the highly realistic images we had seen on Crete: two young boys boxing; girls collecting saffron from flowers; a teenage boy carrying fish. The murals gave us an incredibly vivid picture of daily life: bulls chasing onlookers; swifts swooping through the sky; butterflies darting between fruit blossoms. We looked closer. There were more exotic images: lions leaping on deer; a herd of delicate oryx poised to flee. The wall paintings show animals foreign to Thera – African lions and monkeys, Arabian oryx. Where did these exotic influences come from? A tobacco beetle indigenous to America has even been found buried in pre-1450 BC volcanic ash.

As we moved on through the exhibition, we saw that it wasn't just the art on Thera that reminded us of ancient Crete. The link between the people of the two islands must have been remarkably close. Many of the ordinary domestic objects in the museum were exact copies of what we had found on Crete. Ah, you might say, a soup bowl is a soup bowl, in whatever country. Everyday objects are similar because they have the same function. But the experts specialising in both art and archaeology see a connection too: they even talk about the development of a 'kitchen kit' that became standard here – and on the other islands round about.[3]

While their artistic styles seemed remarkably similar, the resemblance between the societies' basic artefacts, such as fastenings and pins, was striking. To our eyes at least, the Therans' achievements in architecture and engineering were almost carbon copies of what we'd found on the larger island. As in Crete, some of the fine, ashlar-faced houses on Thera were three-storey mansions. Meanwhile, the scale and sophistication of the Bronze Age metals technology on Thera was, as on Crete, astounding.

At all three sites – Knossos, Phaestos and, as it now looked to my eyes, on ancient Thera – there had been marble-floored palaces of incredible luxury. Space, light, freedom: this was not what we normally regard as prehistoric living – this was hot and cold running heaven. The distinctive architectural features of ancient Crete – the light-wells, terraces, central courts, sunken baths and terraced and porticoed gardens – were here too. The open verandas and grand staircases must have made life on this already warm and welcoming island gracious and comfortable. They had fresh, running water, both hot and cold, and they even had a form of air conditioning.

According to one expert, Malcolm H. Wiener:

> When all of the categories of evidence are considered together in the context of Minoan power, wealth, population, trade networks and neopalatial expansion both within Crete and abroad, a major presence of Minoans and descendants of Minoans on Thera seems certain. They may have begun to arrive as individuals or an enclave in the protopalatial period, their number growing subsequently through further immigration and intermarriage.[4]

Still, I didn't have time fully to take in the startling implications of this idea. Because the most amazing fresco of all lay ahead of us. As we entered the room devoted to the West House, Room 5, Marcella and I could hardly believe the evidence of our eyes.

|||||||||||||||||||

A fleet of ships was just returning to harbour. Preserved for thousands of years beneath mounds of volcanic tufa, the images of the homecoming flotilla were largely intact and the colours as warm and glowing as if they'd been painted just a few days before.

Shouting to their friends, teenage boys rush through the town gate, along a narrow strip of land between the sea and the town walls. Women – perhaps mothers and wives, one lady with her young boy next to her – peek out of windows and balconies. Fishermen clamber up the slope from the beach, trying to reach the top of the hill and get the first view of the fleet already filling the harbour. The homecoming of the fleet is the climactic end of a story. What was the full story, I wondered? To view the magnificient Thera flotilla fresco please go to the first colour plate section.

Before us was what amounted to a stolen moment in time: a snapshot of an entire Bronze Age fleet, looking just as it did when it sailed into harbour 3,500 years ago. These must be by far the oldest images of European ships in existence, I thought.

Thinking about the ships took me straight back to Phaestos. Urgently, I scrabbled back through the guidebook. Minoan pottery – particularly the extraordinary black, cream and orange pots known as Kamares ware – has been found in various Egyptian excavation sites dated as early as Egypt's 12th Dynasty.

The Minoan pottery must have been transported to Egypt in ships like these, pictured right in front of us. These frescoes had been on the walls of an astonishing mansion found by Marinatos. The excavators thought the house had belonged to an admiral. If the Minoans had really possessed a well-developed naval command structure, to the extent of having ranks and leaders, then just how well travelled must they have been?

Could a Bronze Age people of 2700 to 1400 BC really have

THE FLEET FRESCO: POSITION OF THE SHIPS

These beautiful paintings, known as the 'Miniature Frescoes', show a spectacular procession of vessels moving between two harbour towns. The ships have figureheads at their prows: they look as if they are carved and painted as leopards and lions, but this could be the actual animal skins. There are garlands draped around these victorious ships, and the townsfolk are excited and jubilant. Every detail is so real and lifelike that I began to speculate that they *were* real; that they were a historical record, and that therefore, it could be possible to find the actual harbours they depict.

I spent 10 years as a navigator and then Captain of submarines, taking periscope photographs and making maps from them. I applied that knowledge to the area of coast west of Akrotiri (on Santorini).

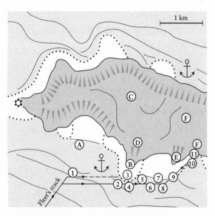

Linking the Fresco to the coast

The map based on Imray chart G33 has been marked, to show

A. Red Beach.

B. The protruding cape at the East of Red Beach.

C. The high hill behind Red Beach.

D. Two roads leading inland from Cape B.

E. Three roads leading inland to the Base (now being excavated).

F. The excavations at Akrotiri.

The map has been 'anchored' on the lighthouse at 36°21'30"N and 25°21'30"E.

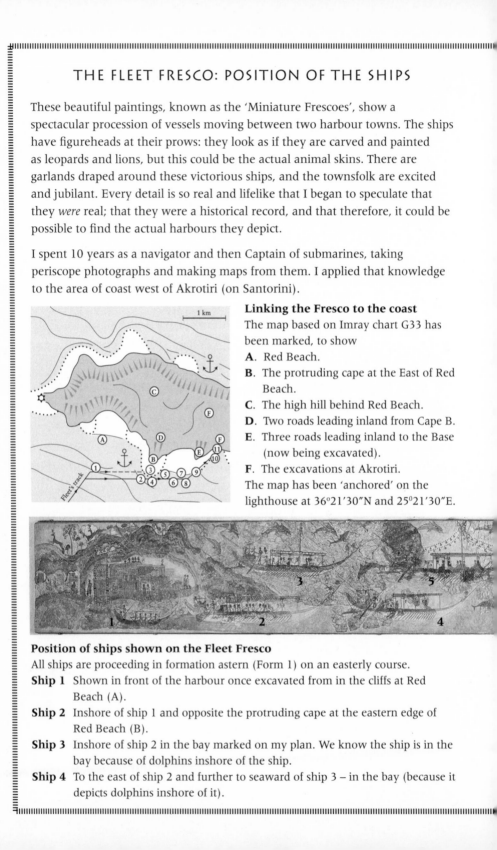

Position of ships shown on the Fleet Fresco

All ships are proceeding in formation astern (Form 1) on an easterly course.

Ship 1 Shown in front of the harbour once excavated from in the cliffs at Red Beach (A).

Ship 2 Inshore of ship 1 and opposite the protruding cape at the eastern edge of Red Beach (B).

Ship 3 Inshore of ship 2 in the bay marked on my plan. We know the ship is in the bay because of dolphins inshore of the ship.

Ship 4 To the east of ship 2 and further to seaward of ship 3 – in the bay (because it depicts dolphins inshore of it).

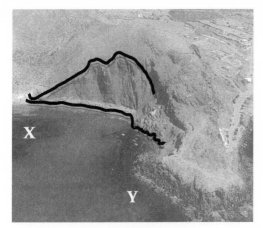

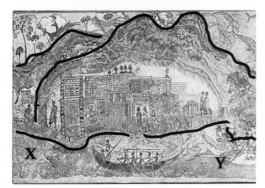

Tying the ships on the fresco to the coastline

The fresco is drawn in 3 tiers – tier 3 can be ignored as it is a river. The top tier shows the western part of the coast with ships 1–5. The middle tier shows ships 6–11 crossing the bay to reach harbour. The ships are numbered on the Imray chart.

Everything now becomes clear. The fleet sighted the high hill behind Red Beach (**C**) 30 miles out in the Mediterranean and steered towards it. When they reached shallow water they turned eastwards for home. Ships 10 and 11 are shown in harbour, ships 2 to 9 en route to home.

Note: the Anchorage signs are shown on the Imray chart. The harbour was presumably chosen (**F**) because deep water reaches there (7m^3).

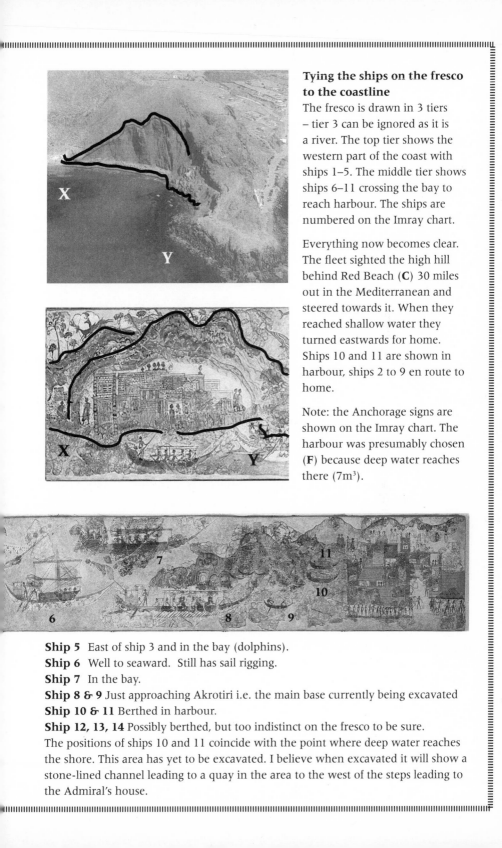

Ship 5 East of ship 3 and in the bay (dolphins).
Ship 6 Well to seaward. Still has sail rigging.
Ship 7 In the bay.
Ship 8 & 9 Just approaching Akrotiri i.e. the main base currently being excavated
Ship 10 & 11 Berthed in harbour.
Ship 12, 13, 14 Possibly berthed, but too indistinct on the fresco to be sure.
The positions of ships 10 and 11 coincide with the point where deep water reaches the shore. This area has yet to be excavated. I believe when excavated it will show a stone-lined channel leading to a quay in the area to the west of the steps leading to the Admiral's house.

constructed the world's first ocean-going vessels, loaded them with precious goods and started plying the world's first international trading routes? What if the centres of Knossos, Phaestos and Thera, island paradises separated only by a few short miles of sea, had been the central hubs of a much bigger sailing nation?

If the Minoans had mastered the supreme skill of shipbuilding they could have spread their growing influence by sailing these very same ships that were depicted in front of us now. The idea of a trading fleet or even a naval power made absolute sense: nothing less would fuel the sophistication and glamour of the Minoan culture, whose richness could still be seen on these walls. Nothing other than thriving international trade could have provided the extreme luxury and wealth which these people had obviously enjoyed.

llllllllllllllllllllll

To my eyes, it certainly seemed possible: at least three of the eight vessels in the 'admiral's fresco' looked as if they could cross an ocean. There is so much realistic information on the frieze that you can even see the number of oars they had, work out the type and efficiency of the sails and estimate the sailing capacity of the ships. The size of one particular ship with a stern cabin suggests that it was the admiral's flagship – which explains its prominent position on the walls of his beautiful house.

Thousands of years ago, Room 5 was probably once a graceful reception room. Before disaster struck, it had had picture windows overlooking the sea and an impressive, wide staircase. I decided to rename Room 5 'the ballroom', as an *aide mémoire*. You can imagine the scene: men and women celebrating an arrival, a victory, or some very special occasion. The admiral's guests are admiring the fleet as shown on the fresco while they gather for drinks at dusk, before dinner. Perhaps they could see the real-life ships sitting at anchor in the bay below them, home after a long, adventurous voyage.

The superb fresco ran along the top of all four walls of the ballroom. Once in the room, guests would have been able to read the story of a voyage to distant and exotic lands – moving anticlockwise, rather than our habitual left to right – and the sailors' victorious return to their home port, where the ships are joyfully greeted by the heroes' families and by the townspeople of Thera.

This was extraordinary, like diving through a porthole into a lost age. By pure chance I was on the trail of an untold and long-forgotten story. I began taking notes. The length, width and draught of the ships – and hence their seagoing capability – can be easily calculated, as well as the weight and volume of the cargo which could be carried. In short, the frescoes might even be able to tell us how far the ships could have sailed and in what weather.

To start analysing all this, I numbered each ship, starting with one on the far left and ending with 9, 10 and 11 – seen on the right-hand side of the picture, nearest to the admiral's house. The ships varied in size. Numbers 4, 5, 7 and 8 are the longest, and 5 and 8 have the largest number of oars: twenty-six a side. Number 6 has the most sophisticated sailing capability.

In fact, excavated Minoan seals[5] show that their ships had significant sophistication, including masts and sails, from at least 3000 BC. Inscribed clay roundels found at Chania, on the western coast of Crete, show the same symbolism for ships as the frescoes did – a hooked prow, a kind of cabin astern.

The small 'cabins' on numbers 1, 2, 3, 4, 5, 6 and 8 were in fact awnings, to shade passengers from the fierce sun beating down on them at sea. They look much like the decorated pavilions that knights in medieval Europe took to war. Amazingly, the remains of one of these cloth coverings had actually been found on the excavation – complete with the manorial arms and crests of a captain and an admiral. Yet more evidence of a highly controlled and well-planned naval infrastructure. The cabin was portable, as were the central square awnings. When they were at sail, the ships would have looked most like ship number 6 – which is dominated by a large, square-rigged sail on a central mast.

Ship number 6, I decided, was the most interesting. It had ten guy ropes controlling the sails: sailors call them sheets.

Using the sheets, a sailor could lower and hoist sails; reduce sail area (by furling); and alter the shape of the sail to get maximum forward thrust. The ropes were controlled using a brass adjuster at the masthead; I later found an early example of a similar system in a museum in Athens. This system of ten ropes allowed you to adjust the sail to maximise sail-power in any wind – full sail mounted crosswise when in light stern winds; lightly furled sail with the curvature narrowed when sailing into the wind; completely furled sail in a squall or in no wind, when the oarsmen could take over.

Each oarsman would need a space of at least 76 centimetres (2 feet 6 inches) in which to row, but 91 centimetres (3 feet) was comfortable and 107 centimetres (3 feet 6 inches) allowed for fast rowing. So twenty-six rowers needed a distance of 24 metres (78 feet) – ships 5 and 8. The oarsmen occupied half of the length of the ship, from the stern to the end of the prow, so this would make the ship 47.5 metres (156 feet) long, with a hull length of about 36.5 metres (120 feet). These were big seagoing ships, about the same size as the *Golden Hind* Sir Francis Drake sailed in Elizabethan times.

Thoroughly transfixed, I examined the fresco more closely. The vessels have several unique features, as far as I know. The first is the projection at the rear shown in ships 2, 3, 4, 5 and 8.

I thought the strange projections could only be hydroplanes, similar to submarine hydroplanes at the stern of a submarine. Stern hydroplanes are horizontal rudders. They adjust the position of the stern. The one on ship 8 would raise the stern and hence lower the bow, altering the configuration of the ship. In a stern sea, or following wind, it could be used in this position; in a head sea or wind it would be reversed, to lower the stern and raise the bow. You could probably shift the crew forward and aft to achieve the same result.

The ships shared another unique feature in this scene. You can see some 'lighter than air' objects floating above the bows of ships 2, 3, 4, 5 and 8. Close up, these objects must have put considerable upward force on the bows, because they are attached to the bowsprit by a thick piece of solid wood. It would seem that these vertical 'hoists' complemented the stern hydroplanes, giving additional lift when the hydroplanes helped to lower the stern (the hydroplanes are reversed from their position in ship 8). How these vertical hoists worked is a complete mystery to me – but I can think of no other explanation for them being there. Mere decorations would not have needed to be secured to the bowsprit with such thick wood.

As far as I know, these two features are unique to Minoan ships.

The third unique aspect of this fleet is aesthetic: the beautiful paintings on the sides of the hull. Ship 5 has lions and dragons; ship 6 sports doves. These extremely sophisticated ocean-going ships were capable of sailing in most weather conditions and were able to adjust both sail and hull configuration. What technology! And just like the airlines or train companies of today – such as BA or Virgin Atlantic – the Minoans branded their beautiful, world-beating ships with their own unique identity, born of their extraordinary talent in the fine arts.

3

THE SEARCH FOR THE
MINOAN NAVAL BASE

I needed to see for myself if Thera had once had a harbour deep
enough for seagoing ships. Where would an 'Admiralty House' be
built in relation to this harbour? Having visited naval houses all
over the world, I know admirals often wish to live near their
principal port, so they can greet ships when they return and enter-
tain captains, officers and men. As a symbol of status, 'Admiralty
House' is almost always built on a hill overlooking the port, like
Mount Wise near Plymouth.

We needed to narrow the search. Any captain or admiral choos-
ing a harbour would require shelter from the prevailing wind and
sea. The prevailing wind on Thera is our old friend the Meltemi,
known by the ancient Greeks as the Etesian winds. The system
results from the high pressure system (>1025) frequently lying over
Hungary and the Balkans and the relatively low pressure system
(<1010) over Turkey.

This helped me pinpoint our search. The port would almost
certainly have been built on a southern shore – either between Cape
Agios Nikolaos and the foot of Mount Prophet Elias (see map), or
on the south coast between Cape Akrotiri and Vlichada.

The ballroom frescoes show the land falling away to the sea,
in a dramatic panorama of the coastline. My guidebook included
reconstructions of maps of the coastline as it would have been in
1600 BC. The beach extended about 300 metres (980 feet) further
to seaward than today, but the shape of the coastline remained
broadly the same: though the admiral would have seen rather more
of the coast and its anchorage than he could today.

Thanks to millennia of volcanic activity, Santorini must be one of the few places in the world that has beaches in three colours – with white, red and black sand. This headland has the strangest coastline I have ever seen. One of the beaches is called Kokkini, or Red Beach – and indeed everything is red, from the terracotta-coloured cliffs, the red sand and red pebbles to the sometimes red tourists, in places. But the water is deep, cool and crystal clear, shading from aquamarine through turquoise to cobalt and then black.

Thinking about it on a practical level, as a seaman, the most likely place for a port would be between Red Beach, 1,000 metres (3,300 feet) west of where the admiral's house was excavated, and White Beach, which is 2,000 metres (6,600 feet) further west.

Then it suddenly struck me – imagine if the admiral's guests were actually able to see the ships: not just on the fresco, but at sea? Supposing the painter of the frescoes was in fact painting the fleet just as it was – moored right in front of the admiral's house? It made perfect sense. Immediately we decided to sail out along the short stretch of coast that looked like the scene in the fresco. We racketed around the harbour until we found someone with a boat and then we set off, the wind in our faces and camera at the ready.

In 1600 BC there were a number of volcanoes affecting this island: La Thirasia, Megalo Vouno, and Mikros Prophet Elias, in the north. These would rule out a base on the sheltered north coast of the island. The port must have been on the south coast.

The sailors would want to be as far away from danger as possible – and all the evidence so far is that this strategy worked. No shipwrecks caused by the volcano have yet been found.

This was a trip to remember: the salt spray flying high as the cutter bounced and plunged its way along the spectacular coastline. The next major requirement of a port is for drinking water, from streams or rivers entering the sea, to stock ships before a voyage. On the south coast between White Beach and Vlichada, I counted thirteen streams. That would be why the merchants had chosen Akrotiri as the main port – to service the fleet's need for water.

We trained our telescopes and zoom lenses on the red cliffs behind the beach. As I framed the image in the lens, the link became unmistakable. Even today, the contours of the land match up with the fresco. The long-buried image Marinatos had uncovered wasn't just a pretty picture: it is in effect a map. You can see the three peaks of the island quite clearly. The only major difference is the shape of a central waterway, at the middle of the fresco.

On the right-hand side of the fresco were the houses now being excavated, not least the admiral's mansion. A broad stairway is shown leading up from the beach to the house. Then, proceeding westwards, tracking left across the painting, you can see a large building with triangular windows: it might have been a type of barracks, interesting to us as we look for evidence of a well-manned fleet, or, less helpfully, a prison. The site has yet to be excavated.

Tracing a line further west along the image of the fleet and the harbour we had in the guidebook, we came to the distinctive, pyramid-shaped hill behind Red Beach.

What we also saw from seaward, to our surprise, was a series of caves along the base of the red cliffs, one or two of them looking as if they were inhabited. We waded ashore to discover that this whole stretch of coast has a string of such caverns, carved out of the volcanic tephra. Later, at the hotel, Spiros the owner told us his grandfather still keeps his fishing boats in such a cave and that many people live in what's known as *skaptas*; long houses carved out of the tephra. They are easy enough to dig, but strong enough not to collapse. Many of them have been turned into restaurants. Starving, and with just a touch of sun and sea burn, we dined in 'Cave Nicholas' that evening, a cavernous room dug 12 metres (40 feet) into the cliffs, which yielded an excellent fish dinner.

Finding an entire lost city – many thousands of years old – was simply breathtaking. The drama of the story, along with the suddenness of ancient Thera's disappearance and its equally sudden discovery, has led many people to speculate that this was the long-lost civilisation of 'Atlantis'. I knew the tale well: it was a highly popular story when I was younger. A fabulous city had

been suddenly drowned underneath the sea. This was the gods' punishment, apparently, for its people's arrogance and hubris. I just dismissed this idea as nonsense. There was huge drama to Thera's destruction, true. It was terribly poignant that at the very height of its brilliance this civilisation's bright flame had been extinguished in a moment. Yet this wasn't a tragedy on the scale of Pompeii, it seemed. The archaeologists were convinced that most of Thera's citizens had escaped the ravages of the volcano. During Marinatos' excavations no actual bodies were found and there are many signs of preparation for the disaster: expressive details, such as food storage jars being left with a covering to protect against roof falls. Most valuables appear to have been removed, suggesting that the people of Thera had been given plenty of time in which to prepare for their escape from the volcano.

The idea of a lost civilisation is fascinating, but in the case of 'Atlantis' the whole concept has become the province of crackpots and charlatans. This puzzling story, and the enigma behind it, has inspired everything from poetry and science fiction to Hollywood films. Where was Atlantis? No one knew, but lots of people had an opinion. Everyone, from the 1920s occultist Rudolf Steiner to SS Reichsführer Heinrich Himmler, embroidered upon the myth. The Atlanteans have become all things to all men, accused of being anything from a race of Nordic supermen to intergalactic spacemen. Three thousand years ago ... it was astonishing to think of being plunged so far back in time that there were no books, no libraries, no contemporary written records – or at least ones that we could understand. As we read in one of our guidebooks, the only early record of Atlantis was a word of mouth account. It was later picked up and written down by a certain young Greek philosopher who was writing a theoretical 'dialogue', the details of which may or may not be true. It stretched my credulity that Plato's voice, reaching out over the wine-dark seas of time, could possibly hold any truth.

> In a single day and night of misfortune, all your warlike men in a body sank into the earth, and the island of Atlantis in like manner disappeared – into the depths of the sea.[6]

Plato's original texts, which had caused all the furore about the mysterious land of Atlantis, were in fact two 'dialogues' called *Timaeus* and *Critias*. Born 423/427 BC, Plato was the second among a great trio of ancient Greek philosophers, along with Socrates and Aristotle. *Timaeus* is concerned with the creation of the universe. *Critias* is incomplete, breaking off suddenly in the passage which concerns Atlantis.

Translations of the texts seemed to suggest that the ancient metropolis and the Royal City were separate entities, which I could see had a strong resemblance to the relationship between Crete and Thera. The main city, he said, lay on a circular island about 12 miles (19 kilometres) wide. The Royal City, meanwhile, was situated on a rectangular-shaped island. So Plato's Atlantis was certainly two islands and possibly more. There are plenty of islands to choose from in the Med. Yet given that Plato was writing a morality tale, it didn't necessarily follow that the details of his fable had to be true. His motive was not to tell future generations about Atlantis; his concern was to discuss philosophy and the failings of human nature. But his account was striking.

> Now in this island of Atlantis there existed a confederation of kings of great and marvellous power, who held sway over the island, and over many other islands also.[7]

I still dismissed the whole thing from my mind as nonsense.

As we observed in the *skapta* restaurant that evening, for the ancient Therans, murals must have been a kind of photography – a documentary as well as a decorative art form. The ships are shown proudly anchored in full view of the admiral's house, almost as if they were a memorial to him.

To my mind the harbour at Thera is actually far more impressive than those at either Amnisos or Kommos, the ports on Crete that served Knossos and Phaestos. One could imagine the Minoans from Crete first joining forces with Thera to face the enormous challenges posed by the Bronze Age. This large, deep-water port had the potential to be truly international.

The paintings tell the tale: of Egyptians in white robes, Africans with black curly hair and intriguing thin prisoners with red skins, who look northern European. They were documentary proof that there was a huge level of trade going on in the Aegean, before the beginning of recorded history.

In the Archaeological Museum of Heraklion there was plenty of supporting evidence: for instance, we'd found a staggering stone axe, skilfully sculpted in the unmistakable form of a panther (see first colour plate section). Panthers come from South America. Not the Mediterranean. One thing above all kept coming back into my thoughts: the tiny beetle that had been found in the volcanic ash. It came from a totally unexpected place: North America.

4

RETURN TO PHAESTOS

We returned to Crete with new eyes. Strolling through the market in Akrotiri, we'd seen that on this one remarkable island nature had provided almost everything that ancient man might have needed. It looked as if the Minoan Crete–Thera complex had been an international marketplace, exporting foodstuffs like olives and olive oil, figs and saffron, while importing other goods in return. Perhaps it was also one of the world's first true cosmopolitan societies? We counted 28 different varieties of fish in the market and sampled what may be the world's finest olives, plump as plums and twice as tasty.

Olives formed the basis of early commercial life on the island: the oil, say the experts, has been exported across the Mediterranean for millennia. As Dr J. Boardman wrote:

> Crete started to cultivate olives from the wild specimens early in the Bronze Age around 3000 BC. At Myrtos, a port on the south east coast of Crete, vats from this era for separating olive oil have been found. The early method was to crush the fruit, then plunge it into hot water which separated the oil before skimming it off. From olive stones found at this site it seems Crete may have been the first Mediterranean country to cultivate olives.[8]

The island certainly has the ideal geography and climate to do so. Olives are a hardy fruit, able to withstand the long, dry heat of a Cretan summer with its months of drought. A cold spell in winter usually plumps up the fruit for the principal harvest. The olive harvest comes conveniently after those of wheat and grapes, and before the sowing season – November to January. Life begins at 40 for the olive tree, after which farmers can hope for 50 kilograms of oil every second year. Provided they are tended and pruned to prevent them reverting to the wild, olive trees can live for centuries. Farmers plant for their grandchildren and those children plant for their own descendants, in a cycle that has gone on for thousands of years and is one of the first and best examples of sustainable agriculture.

One peculiarity is that heavy crops come in alternate years. This two-year cycle seems to be the same the world over. On Minoan Crete, the plentiful olive harvest every other year meant that Minoan palace bureaucracies had to come up with complicated arrangements that included extensive storage facilities and detailed planning to distribute the oil. In ancient Greece merchants booked olive presses two years in advance, just as farmers today book combine harvesters.[9] Dr Boardman cites evidence of huge quantities of olive oil being held in stock in the great storehouses and vats of the Cretan palaces – the West Magazine of Knossos Palace could store 16,000 gallons. Cretan Linear B tablets record details of the oil that was stored and distributed in the 14th and 13th centuries BC.[10]

Just how crucial this product was to the ancient world is demonstrated by the way in which the burning of olive trees eventually became a tactic of warfare, as evidenced in accounts of the wars between Athens and Sparta. Olive oil had numerous uses, quite apart from enhancing the flavour of food. It was used as an offering to the gods, as well as forming the basis of the Minoan cosmetic industry. Vases with a special shape (*pelike*) were used in the cosmetics trade and sales were carried out with special dippers and funnels. The palace of Knossos has records of hundreds of olive trees being used to create perfumed oil. The cycle of olive production, from harvest to distribution, is depicted on Athenian vases,[11] which show men beating the trees with long sticks to get the olives to fall, as they still do today. A statement on one such vase says: 'Oh, Father Zeus, may I get rich'.

Heading once again in the direction of Phaestos, this time we drove over the majestic Psiloritis mountain range, the highest point in Crete, where the olive trees vanished in favour of dark green pines, juniper and evergreen oak. Up until quite recently, the shepherds in these hills built beehive-shaped stone huts called *mitata*; seeing one, I could not help thinking of the prehistoric *tholos* tombs built by the warlike Mycenaeans, who were later arrivals on this island. The *mitata*'s traditional, stove-like shape transcends millennia.

We stopped at a mountain village for Greek coffee. The elderly innkeeper invited us to view the courtyard where there was a wood oven and we shared the space with a cluster of cats nestling for warmth. The enticing smell of roast meat and stuffed eggplants was coming from a large earthenware dish inside the oven. Naturally, we were unable to resist. This turned out to be the best kid I have ever eaten. It was first marinated in olive oil and lemons and then finished by cooking slowly with wine, basil, thyme and wild green vegetables – absolutely sumptuous. Crete's soil is so fertile that fennel, leeks and other wild leafy vegetables can simply be gathered at the side of the road. The island has the largest number of herbs growing wild in the world. Over the next few days I investigated the food that Cretans ate thousands of years ago, which has been analysed from remains found in their cooking pots. They ate kid cooked in just the same style as the dish we enjoyed. They also ate salads – of lettuce, onions, garlic and celery, again flavoured with olive oil and a variety of fruits and nuts; almonds and pistachios were popular. They drank Retsina (resinated wine) and beer brewed from barley. Like the superb Cretan olive oil, these products would soon become wildly popular abroad.

As the first view of the ancient palace emerged through the trees I decided to stop for a minute. We got out of the hire car and looked down on the velvet carpet of crops below us, stretching away like the Garden of Eden. We were at an altitude of about 120 metres (395 feet) above sea level. At this height, what were in fact the walls, pools and stairs of the palace looked like hieroglyphic markings. Rivers, we now realised, flowed all around the palace and then tumbled downhill towards the Minoan harbour. Phaestos, light and flawless under the azure sky, was at the centre of a landscape made for adventure.

As we drove off again we saw a sign pointing down a dusty track. I recognised the name. This was the remarkable Kamares cave I had read about; an extraordinarily deep shaft cave which archaeologists think had a sacred or cult religious role in the very early days of Crete.

The pottery told us loud and clearly that the Minoans had traded much more than foodstuffs and olive oil. The Kamares designs are dramatic, a modern-looking black and red, and the pottery was first excavated here in the early 1900s. I'd learned by now that it had been highly prized across the entire Mediterranean. It has been found across the Levant and Mesopotamia, from Hazor and Ashkelon in Israel to Beirut and Byblos in Lebanon and the ancient Canaanite city of Ugarit, near what is now the sea-town of Ras Shamra in modern-day Syria. Judging by the finds in Egyptian tombs and elsewhere across the region, the Minoan skill in art seems to have given the Minoans of ancient Crete a free pass to the glamour, science and civilisation of the two most advanced cultures of the Early Bronze Age: Mesopotamia and Egypt. Minoan creativity must have been a pearl beyond price. Minoan creative output was precious, perhaps almost holy, like the production of silk was to the Chinese.

We arrived at another one of those strange Cretan road signs that don't appear to refer to your real destination but end at a blank wall, or in this case a blind bend in the road. At first we shuffled about a bit, not knowing where to go in this remote spot. Then we realised that to get to the mouth of the cave – a steep, knee-taxing ascent – you had to turn off the road and climb on foot. You are rewarded at the top by breathtaking views through the trees to the plain of Messara and the Libyan Sea.

To visit the cave, you scramble through the gaping mouth of the entrance and down the giant, slippery boulder steps – built surely for monsters – scrabbling and sliding over the scree. The plunging path leads into a huge vaulted chamber, nearly 100 metres (330 feet) long. The atmosphere is intense and other-worldly; dark and alive with the spirit of an ancient age. The slope downwards is steep, some 40 metres down to a wide boulder-strewn floor, then it turns into a narrow, twisting passage. A second, much smaller inner chamber slopes down another 10 metres. You move in complete darkness until you reach an unmistakable, rather foetid smell: water. No wonder this evocative place remained embedded in folk memory. Deep in the bowels of the mountain, it is easy to see how a cave like this might have inspired the myth of the Minotaur's labyrinth. In myth, the young hero Theseus, blindfolded, had to try to find his way out of a place that you imagine to be as clammy and slippery as this: a labyrinth that was in truth a prison. There he would have had to confront the half-man, half-beast.

When the cave was excavated in 1913, hundreds of shards of Kamares pottery were found, together with animal bones, fragments of terracotta animal figurines and tools in both stone and bone, as well as six iron spearheads that were probably post-Minoan. Going by the differently dated finds, experts believe that the Minoans used the cave during the whole of the Bronze Age era (c.3000–1100 BC). The cavern is so huge, so daunting and so easy to defend that in later times it had a history of being used by whole villages to escape from danger. Yet the sheer amount of ancient Minoan Kamares pottery found inside it suggests that this wasn't simply a refuge: it had been used over the centuries for long-forgotten sacred rites. Caves seem to have been highly significant to the Minoans, in religious terms. Professor Marinatos found scores of precious bronze weapons hidden in the Arkalochori cave and hundreds of cult items have been found here at Kamares, from figurines to double-headed axes.

An extensive network of roads links the palaces at Palaikastro and Kato Zakros with the southeastern part of the island. I wondered if these ancient palaces had been among the earliest known city states, run like independent fiefdoms with dependent villages, much like medieval European towns. It seemed not. The engineers who built Crete's prehistoric roads had set up way stations and watch-towers, suggesting that lots of valuables had perhaps been moved along the roads. The Minoans had wanted to protect the goods travelling along them. Yet if each palace had been independent – and especially if those palaces were each other's rivals for political dominance – that degree of co-operation was unlikely. So in the Minoans we were talking about an amazingly sophisticated and developed people. It seemed they understood the value of an inter-connected society. They had cared for and looked out for each other; they had worked together for the common good.

We reached Phaestos, or Festos as the locals call it, in the heat of the noon, and waited in a tiny village café for the sun to lose its strength while I plunged back into my research. In the 14th century BC, said Professor Alexiou, the bounty of Crete – its skilled metal-work, olive oil, pottery, saffron and so on – was exchanged as gifts between eastern Mediterranean rulers. In return, the Egyptians sent exotica: gold, ivory, cloth and stone vessels containing perfumes.

The wealth of pottery, sculpture and jewellery that had been found on Crete was so old that no one could accurately date it, according to Professor Alexiou. So many ancient Minoan artefacts are in Egypt that experts are best able to date Cretan finds by comparing them to Egyptian ones, whose chronology is better understood. According to Professor Alexiou:

> The absolute date in years of the various Minoan periods is based on synchronism with ancient Egypt, where the chronology is adequately known thanks to the survival of inscriptions. Thus the [Cretan] Proto Palatial Period [2000–1700 BC] is thought to be roughly contemporary with the [Egyptian] XIIth dynasty [1991–1783 BC] because fragments of [Cretan] Kamares pottery attributed to Middle Minoan II [c.1800 BC] have been found at Kahun in Egypt in the habitation refuse of a settlement found in the occasion of the erection of the royal pyramids of this [XIIth: 1991–1783 BC] dynasty. One Kamares vase was also found in a contemporary tomb at Abydos [Egypt – Valley of the Kings]. The beginning of the Neo Palatial

period [Crete – 1700 BC] must coincide with the Hyksos epoch [1640–1550] since the lid of a stone vessel bearing the cartouche of the Hyksos Pharaoh Khyan was discovered in Middle Minoan III [c.1700–1600 BC] levels at Knossos [Crete]. Equally the subsequent Neo Palatial Cretan period [1700–1400 BC] falls within the chronological limits of the new kingdom with particular reference to the [Egyptian] XVIII dynasty [1550–1307 BC]: an alabaster amphora with the cartouche of Tuthmosis III [1479–1425 BC] was found in the final palatial period at Katsaba [Crete] ....[12]

Exploring the palace was not our mission today. We planned to walk south from the palace precincts to the prehistoric port of Kommos. What may even be the world's first paved road lay just a few hundred metres away from where we were sitting. Venturing out into Pitsidia village at about 3.00 p.m, we headed for the pretty, crescent-shaped beach of Matala. We would find the excavation of the port somewhere along here. Finally, we found a half-obscured sign on the right, next to a small supermarket. Tamarisk trees offered some pleasant shade along the way.

We marvelled at the road's perfectly fashioned stonework. The finely cut ashlar stone bed even has a slight camber: water would still drain off it easily after winter storms. The Minoan town was spread out over a low hilltop to the north and on the hillside south of it. Archaeologists say that Kommos' larger buildings were built c. 1450–1200 BC: that is, in the Minoan Neopalatial and Postpalatial periods.

It was remarkable to look over the low walls of the ancient town and think of how this quiet spot might once have felt, bustling with sailors making and mending their equipment, perhaps with vendors selling goods and refreshments on the quay.

The University of Toronto has been carrying out excavations here since the 1970s. Most – but not all – of the remains of Kommos consist of a single layer, or course, of cut stone. But the archaeological digs prove that this town had once been a major harbour, with large and imposing houses, grain storage, a central square and some monumental buildings. We could see raised paths and rows of steps, along with what looked like long, open, broad areas that would have been paved with stone, perfect for landing cargo. An enormous palace-like building, called for now J/T, had an extensive colonnade facing on to a court in its centre. It does not seem to have had the spaces for religious ceremony that you would expect in a palace.[13] Its back wall and much of its floor had been covered with vividly coloured and spiral frescoes.

The so-called 'Building P' was also fascinating. It might have been a storage yard for the Minoan fleet's sailing ships, when masts were lowered during the winter non-sailing months. Alternatively it was a vast storage shed, capable of warehousing large amounts of goods, ready for shipping. It had at least four 5.60 metres – (18 feet) – wide east–west galleries, which had no signs of closure at the west, where they face the sea. A broken Minoan limestone anchor was found in one of the long galleries, the kind used by large seagoing ships. Analyses have found that the stone for the anchor slab was quarried in Syria. It is pierced by three holes: one for the thick rope linking it to the ship and two smaller ones for the pointed wooden stakes that would hold the anchor to the sea floor. Having found huge storage jars here in the 1920s, Arthur Evans speculated that this could have been the 'Customs House' of Crete. He was not far wrong. According to the University of Toronto:

> Our excavations have borne out Evans's suppositions regarding the commercial nature of the site and greatly surpassed our own expectations of what might be found at this beach-side spot. After 25 years of digging, Kommos is revealed as a major harbour, with monumental Minoan palatial buildings, massive stone storage complexes and a Minoan town (ca. 1800–1200 BC) ... The portable finds, which range from stone anchors to local and imported pottery and sculpture, speak of the seagoing interests and mercantile nature of the place. Vessels from Cyprus, Egypt, and Sardinia indicate the

sphere of trade contacts enjoyed by the citizens of Bronze Age Kommos.[14]

As we walked around the sea port's perimeter we realised that its once sturdy houses and storerooms would have withstood the roughest gales. They would have had to. With the strong northwest trade winds in your face you could almost imagine this was a long-lost Cornish fishing village, lying semi-ruined in a welter of ancient weathered stone. Rolling in from the great width of open sea between Crete and North Africa, the waves are impressive and strong; totally unlike the usual soothing calm of the Mediterranean.

What these ancient walls wouldn't have withstood was a tsunami. Kommos, facing the Libyan Sea, hadn't had to, which is why so much of it has survived.

Most experts now agree that a massive volcanic eruption had destroyed Thera c.1450 BC. The Theran explosion was one of the largest volcanic eruptions in about 20,000 years. And the titanic amount of lava pouring into the sea from the volcano, in turn, had triggered a gigantic tsunami. Racing across the sea between Thera and Crete, faster than an express train, the tidal wave had hit the bustling Minoan towns and palaces that for the most part hugged the ancient shore, reducing them to debris in an instant. Some evidence suggests that the wave may at points have been as high as 26 metres, over 85 feet.

The experts quarrelled angrily over the years, but the tsunami theory gradually gained ground. The force of the Santorini volcano, the Dutch geologist Professor van Bemmelen argued, was about 1,000 times that of the H-bomb that had split Bikini Island in half in 1954. And the H-bomb was itself equal to nearly 1,000 atomic bombs of the type that fell on Hiroshima. Many experts disagreed. To some, the disaster came about because of a pyroclastic surge, super-heated steam burning everything in its path. The most straightforward counter argument, though, was that it didn't make any difference that Crete was so near to Santorini/Thera. The damage the volcano caused was tiny.

If that were the case, then what had happened to the Minoans? It was simply a mystery. An unsolvable mystery. But the truly devastating power of a moving wall of water became clear to us all on Boxing Day (December 26) 2004, when Sumatra and the Indian Ocean were hit by a devastating tsunami. Two hundred and thirty thousand people were killed.

With the knowledge gained from the horrors seen at Aceh, Indonesia, on that day in 2004, scientists have since been able to analyse the tsunami phenomenon with the benefit of modern technology and techniques. At various sites on Crete, in residue deposited up to 7 metres (23 feet) higher than sea level, the remains of Minoan plaster, pottery and food were found pulverised together with tiny fossilised sea shells and microscopic marine fauna.

The geologist Professor Hendrik Bruins told the BBC that the shells and pebbles 'can only have been scooped up from the sea-bed by a powerful tsunami, dumping all these materials together in a destructive swoop'. (BBC Timewatch *The wave that destroyed Atlantis*) At one of the largest Minoan settlements, Palaikastro, on the eastern edge of the island, Canadian archaeologist Sandy Mac-Gillivray has found other tell-tale signs of giant waves. Town walls which face the sea are often destroyed, or missing altogether.

'Even though Palaikastro is a port, it stretched hundreds of metres into the hinterland,' he said, 'and it is, in places, at least 15 metres [50 feet] above sea level. This was a big wave.'

Those sea creatures referred to by Professor Bruins live only in really deep water. Today, many think Minoan Crete was hit by the biggest tsunami the world has ever seen. This giant wave must have hit the island at points as high as 30 metres (98 feet) above sea level. As sailors and traders, most of the Minoans' towns were built along the coast, making them especially vulnerable to this disaster.

To my surprise, the geophysical evidence tallied with Plato's account, which I'd read on Thera. Atlantis, he'd said, had been destroyed suddenly, 'In a single day and night'. It was a truly awful event that was followed by a terrible 'darkness'. Sickening plumes of gas and ash follow the eruption of a super-volcano. They would have created a vast black dust cloud that spread poisonously throughout the whole of the Med. The cloud's masking of the sun would, in turn, alter the climate, damaging crops. Today, the Belgian archaeologist Jan Driessen believes that the first wave of awesome seaborne destruction was followed sometime later by famine or epidemic. It was the tail end of a huge spiral of natural disasters.

IIIIIIIIIIIIIIIIIIII

I looked at the ruined warehouses of Kommos, which had been built in stone cut with astounding precision. The fact that stone had been cut at all was the most telling point: it meant that the early Minoans must have had strong, sharp metal saws. And there was only one metal then available to do the job: bronze.

It was that ability to cut hard rock that freed humankind from the Stone Age. To cut cleanly through large blocks of stone, the pre-requisite is hard and sharp tools: bronze saws and chisels, adzes and other tools of the mason's trade. Once equipped with the marvellous implements made of the earth-changing metal alloy, bronze, a whole new world of technology opened up to the human race. It looked to me as if the Cretans had taken full advantage of that technology. They had done this long before mainland Greece and perhaps even before the Egyptians. The thing was, how to prove it.

I suddenly remembered a holiday in Egypt in 1977: our two girls were then still quite small. We'd been reading about the incredible significance of stone to the culture of Egypt. Until the appearance of the world's first known architect, Imhotep, even royal tombs were simply underground rooms topped with mud – they were known as *mastaba*s. Imhotep developed the lump-like *mastaba* into a much more impressive pyramid, built in hewn stone, for the powerful ruler Khafre. That first memorial tomb was simple, certainly by comparison with what was to come in architecturally ambitious Egypt. Nevertheless, it was encased in fine white lime-stone and rose in six steps to 60 metres (197 feet) high.

When we arrived at the Great Pyramid at Giza, our two daughters clamouring for a camel ride, we were amazed at this 146.5 metre tall – (480 feet) – miracle of construction. Napoleon was spellbound when he saw it, the oldest of the remaining ancient wonders of the world – and the only one still standing intact. Our guide had explained the sheer scale of the enterprise. Covering an area of 53,000 square metres (13 acres), the site of the Great Pyramid is large enough to contain the European cathedrals of St Peter's, Florence, Milan, Westminster Abbey and St Paul's together. Building it would have involved putting 800 tons of perfectly cut limestone and granite into place every day for 20 years. Each slab weighed between 2.5 and 50 tons. I'd asked, in admiration, how they had cut these huge blocks of stone with such accuracy. 'Saws,' our guide had said. 'Bronze saws.' Had Egypt imported its bronze tools from Crete?

Minoan Civilization even cited international correspondence with the Minoans, which was documented by the Egyptian pharoahs. All the evidence was that the Minoans' trading missions to Egypt were regular, not one-off events. To refer again to Professor Alexiou:

> There can be no doubt that the Cretan palace-sanctuary unit, with its huge store rooms, played the same central part in economic life, agricultural production and foreign trade as the Temple and Palaces of Egypt and the East. Conclusive evidence for the existence of workshops for stone masons, ivory carvers, makers of faience and seal cutters comes from Knossos and the great Cretan palaces ... Both agricultural produce such as olive oil, wine and saffron, and the accomplished Cretan metalwork pictured in the tombs of the 15th century Egyptian noblemen and described as 'gifts from the leaders of the Keftiu (Cretans) and the islands' were in all likelihood directly exported to Egypt from the Cretan palaces ... The Egyptians in return sent gold, ivory, cloth, stone vessels containing perfumes and chariots: besides monkeys for the [Cretan] palace gardens and Nubians for the royal guard.

We arrived at the beach, where it was very windy, if still sunny, to see the Stone Age cave houses, complete with passages, stone beds and fireplaces, snuggled into the cliffs.

I drove back towards the palace with mounting excitement. It was simply astounding to think of Phaestos guarded by Nubians: of monkeys gambolling about in this Cretan palace 4,000 years ago. Yet the evidence was gathering in front of our very eyes. Our experiences, both here and on Santorini, all added up to one thing. Unthinkable as it seemed at first glance, more and more evidence pointed to the Minoans as the world's earliest maritime traders. They had been consummate sailors and merchants; they had built and sailed some of the world's first seagoing cargo ships; and there were cosmopolitan, sometimes royal, travellers crossing the seas.

I could not let it go. I'd spent the best part of a decade establishing beyond my own doubt that the Chinese had been the world's first global travellers. Now my theory was being turned on its head. In some ways, I was almost galled to find that the Minoans seemed to have sailed to Egypt and perhaps beyond, when my book, *1421*, argued that the first voyages of world discovery were made in AD 1421, over a thousand years after the time of Christ.

We had set out to visit a simple Aegean island. What we'd found instead was the Paris, Hong Kong, New York and London of the ancient Bronze Age, rolled into one. We'd found not an island, but a trading empire.

5

THE ANCIENT SCHOLARS SPEAK

Racially, culturally and in terms of international trade there was no doubt about it: Crete was no ordinary island. Myth already had it that the king of the gods came from Crete: the poet Homer even calls King Minos the 'companion of mighty Zeus . . .'. Feverishly, we began to read more books and hunt through museums. Once you started looking, there were hints everywhere about the existence of a powerful island empire, long since lost to human memory. The ancient historian Thucydides wrote that 'King Minos' captured and colonised the Cyclades, driving out the Carians. And the 1st-century BC historian Diodorus Siculus claimed that five princes sailed from Crete to the Chersonnese peninsula opposite Rhodes, expelled the Carians and then founded five cities.

By now, I had collected lots of research material to read. I settled down on a worn wooden bench, peeled some books out of my rucksack and prepared for a breezy journey through some very ancient history.

A number of academics, I learned, were absolutely convinced that the Minoans ran a series of colonies and that their domination of the seas had led to a *'Pax Minoica'*, or 'Minoan Peace'. The island of Kea, which is much closer to mainland Greece than it is to Crete, was nevertheless a perfect example of a colonial culture that closely mirrored the Minoans.

The British Museum holds collections of Minoan pottery, jewellery and seals found in tombs and ruins spread all over the Mediterranean, in places such as the Gulf of Mirabello, Cyprus, Rhodes and Aigina. Interestingly, some of the jewellery in the

'Aigina Treasure', one of its most famous collections – which was largely made around 1850–1550 BC – proves that there was far-reaching trade abroad (see first colour plate section). The luminous amethysts that the goldsmiths had used could only have come from Egypt; there is also lapis lazuli which experts conclude must have come via trade routes from Afghanistan.

One of the books in my rucksack had been bought second-hand and some of the passages were already marked out for me in yellow marker. The book must have been well loved, because it was thumbed and worn, its pages now curling with the damp of the sea air. I noticed one passage in particular. It read:

> Minos was the first person to organise a navy. He controlled the greater part of what is now called the Hellenic Sea. He ruled over the Cyclades, in most of which he founded the first colonies.[15]

So did this great ruler, Minos, establish a sea-based empire, with colonies, or at least bases, over the whole Mediterranean? The aristocratic historian Thucydides, writing in Greek in the 5th century BC, certainly thought as much. Many modern scholars like Bernard Knapp agree:

> ... Minoan Thalassocracy [that is, control of the sea] during the Middle to late Bronze Age was well entrenched ... Minoan sea power [was maintained], through conquest, transplanted 'commercial colonies', and special trading relationships ... various sites in the Cyclades and Dodecanese and along the western coast of Anatolia [where the Uluburun wreck was found] formed part of a Minoan empire, and Knossos dominated the principal trade networks within the Aegean ... [16]

Moreover, Thucydides' text said: 'Agamemnon ... must have been the most powerful of the rulers of his day ...'.

I had a moment of recognition. The civilisation that took over on Crete, after the tsunami and the decimation of the Minoans' fleet, seems to have been its former colony at Mycenae. The Mycenaeans may well have inherited all that was left of the civilisation on Crete and Thera.

Thucydides went on: 'It was to this empire [of Mycenae] that Agamemnon succeeded, and at the same time he had a stronger navy than any other ruler.'[17]

So Thucydides was writing about the inheritance the Minoans had handed on to posterity: the skill to build ships and to sail them. Crucially, he was writing this nearly 130 years before Plato had written his *Timaeus*.

Suddenly, it all made sense. For many thousands of years the only practical way to travel with any speed was by water, certainly when you were talking about large distances. At the centre of the watery world of the Mediterranean was the island-filled Aegean. It must have been like Heathrow Airport is today; strategically placed between the time zones of east and west. The Aegean had been a transport hub: a vital carrier for settlers, merchants, and diplomats. To get to the more open waters of the west, captains had to navigate through the treacherous passage between the southern tip of the Peloponnese, Cape Malea, and the island of Crete, giving the Minoans of Crete and their satellite islands great power.

They appeared to have possessed remarkable technologies to shore up that power. There was one in particular that I was interested in exploring, because of its implications about trade: the creation of seals.

Vast numbers of ancient seals have been found on Crete. Tiny things, carved out of soft stones, ivory or bone. They speak volumes. Originally the idea of a seal might have been quite simple – to mark out personal property, or to seal jars and amphorae. But the point was, there were so many of them, each tiny piece giving a flash of insight into a vibrant prehistoric culture.

Over the years, archaeologists have found over 6,500 seals at Phaestos alone, stamped with more than 600 different designs. A system on this scale must have been about more than simply protecting private property – the seals had been used as tallies in the economically important business of trade.

According to Professor Alexiou, the doors to the palaces' many storerooms were found securely locked and sealed. If the Minoans were actually exporting[18] their extraordinary Kamares pottery, or their jewellery, or their olive oil, they would need to be organised. They would also need to deter theft. They had to keep lists of what was going where, and to whom. Were those seals needed to mark out goods in huge storerooms, ready to be shipped out across the Minoans' empire? More, did they go further afield, to a luxury, élite clientele?

As I peered through case after case in the island's museums, I was shocked to find some truly astounding designs on some of those ancient seals, particularly on the majestic seal-rings. The tiny, minutely inscribed discs seemed to me at least to be more than just badges of ownership. They were surely a status symbol: a route, perhaps, to sacred knowledge?

I marvelled at the sheer ingenuity of these truly exquisite things. How on earth did they make these tiny, minuscule engravings? The extraordinary answer is that the Minoans had already invented lenses: magnifying glasses, in modern terms. If you'd asked me beforehand, I'd have said that the Englishman Roger Bacon, a lecturer at the University of Oxford, had invented the magnifying glass in the mid 1200s. So much for that.

Such lenses weren't unknown to scholars of the distant past. The Roman Emperor Nero used a wafer-thin, lens-shaped sliver of emerald to correct his short sight. And during the siege of Syracuse in 214 BC the great inventor and mathematician Archimedes burned the attackers' ships using parabolic mirrors. Still, both these examples occurred much later in time. Had the original source of this technology actually been the Minoans?

One lens found on Crete can magnify up to seven times with perfect clarity. That particular lens has been dated to the 5th century BC, although looking at the miraculous level of detail that craftsmen managed to etch on to Minoan seals that are much older, the technology was definitely invented earlier. It seems that the seals – or perhaps the minute designs that had been inscribed on them –

were invested with a special spiritual significance. For instance, many of the lenses had been discovered hidden in a sacred cave, known as 'the Idaion cave'.

Mount Ida is on Crete's highest peak – slightly northwest of Kamares, above the barren plateau of Nida. This is where the mythological goddess Rhea is said to have hidden the infant god Zeus from his terrible father Cronos, who had jealously devoured all his other children. The story goes that to protect the baby Zeus from being eaten, special warriors, known as Kouretes, danced around him with shields and metal weapons, making such a clashing noise that his gluttonous father would not hear him cry. Is this another folk memory of the special power of bronze weapons?

This all adds to the powerful sense of mystery that emanates from the island. There are plenty of caves honeycombing the Psiloritis mountain range, including Sfendoni and Melidoni – along with one that is known to cavers and explorers as the Labyrinth. Much of this cave was used by the German forces to store armaments during the Second World War. They then blew them up during their forced retreat. Like Kamares, the Idaion cave was holy, but in this case it was more strongly associated with the mother goddess and with the deepest layers of pre-Greek myth.[19]

Crete, just as Plato describes Atlantis, was a metalworking civilisation with lush, fertile lands: a land of metal, milk and honey. All this is symbolised by a beautiful little bee brooch we later found in Heraklion's Museum of Antiquities. In Greek mythology, it was Melisseus (literally, the bee-man) who nursed Zeus as a baby. The jewel is a work of great delicacy and devotion; the ancient Atlanteans, or Minoans, were obviously as fond of harvesting honey as they were of making beautiful jewellery (see first colour plate section).

I was still sceptical about the 'Atlantis' theory, but I could quite see why enthusiasts might interpret the Minoan world as Plato's Atlantis. The implied skill with metals technology does, in fact, make 'Atlantis' sound like a Bronze Age culture. In his twin dialogues *Timaeus* and *Critias*, Plato describes his mythical island civilisation as having walls 'covered with brass'.

Just like the bees, the Minoans' wealth was founded on ingenuity and hard work. That incredible wealth, along with the extraordinarily creative skills of its people, must have given Crete a mythical aura, a status and standing that lived on well after the Minoans themselves.

What was certain was that the Cretans had possessed levels of technology we can hardly credit; after the loss of the Minoan civilisation, humanity had to wait for centuries to regain these skills and technologies, eventually rediscovering them in stages many hundreds of years after the birth of Christ. No wonder that, for the ancient Greeks, this magical land had reached the status of myth.

The key to understanding the Minoans seemed, at least on one level, to be the seals. What did they have written on them? Were they navigation instructions? Maps of the stars? Perhaps they were merely ownership details. However they had achieved it, both the skill and the knowledge of the Cretan goldsmiths were astounding. Some of the minutely inscribed seals seemed to show stellar constellations, like Orion. There was a growing conviction in my mind: the Minoans must have understood navigation and they must have used the stars to navigate. Whether or not they became successful world explorers would depend on just how well they could do it.[20]

Back at the hotel, I pored over images of the seal stones and then at the Phaestos Disc, back and front. I counted the pictographs on the disc and wondered if you could arrange the symbols into groups and categorise them. Did it show us something profound? Of course, I realised the disc could simply detail something prosaic – a list, perhaps. Why was the notion of the spiral – seen everywhere on Minoan pottery and jewellery and now here on the maze-like surface of the disc – so important?

In the last century scholars had managed to crack truly fiendish languages like Ugaritic. Even the language that followed Linear A on Crete, the so-called 'Linear B' adopted by the Mycenaeans, had been partly translated: but not this. The Phaestos Disc had eluded everyone who had ever tried to decipher it, while the seals told a story of a civilisation of unparalleled invention and technical ability.

I was fascinated, but my lack of expertise was frustrating.

Our modern world had unearthed ancient tokens like these seals and the Phaestos Disc, tools that could potentially unlock the mysteries of the ancient Minoan world. We held the keys to the puzzle; we just couldn't turn them in the locks.

6

THE MISSING LINK

The Bronze Age is named after a copper alloy. It was a miraculous material. Suddenly, metalworkers could change rock to metal, using the magic of fire. Today, world politics is dominated by areas rich in oil, uranium and knowledge. Many thousands of years ago, metal was as important for the development of wealth and power as energy supply and information are today.

It is not known quite when, or where, some unknown genius discovered that an amalgam of nine parts of copper to one part of tin would produce the metal we now call bronze. The discovery revolutionised the technology of the world. Bronze was sharp-edged, strong and durable. The texture and strength of this material makes it ideal for creating effective weapons and astoundingly flexible tools. This precious metal – for then, it truly was precious – totally transformed the world. It was essential to advancing tech-nology and crucial to the development of civilisation. It gave man a modern material – an alloy – that could be moulded and beaten into any shape.

The new knowledge was a long time spreading. In countries where there were no native copper or tin mines, spears and swords could only be obtained through trade or conquest. They were of fabulous, untold value. The owners of bronze swords, bronze-tipped arrows and bronze shields would have seemed well-nigh invincible, a fact which must lie behind the many ancient legends and myths that concern magic swords and armour.

Suddenly, Bronze Age man possessed hard, corrosion-resistant metallic tools – shovels, axes, chisels and hammers. For the first time in the history of the world, with the free time bought by this helpful technology man was able to create products of pure luxury in large amounts, such as sumptuous jewellery. The Minoans excelled at jewellery production. They also had the sharp saws they could make from bronze to thank for their ability to build durable ships of carved hardwood. These true, all-weather ships could trade these new luxury goods with Egypt and across the Mediterranean.

And let's not forget weapons. While metals such as gold and silver could finance a war, the magical new metal of bronze could win one. The historian Herodotus talks about 'the men of bronze' who 'sold their fighting skills to the pharaohs of Egypt'.[21] We don't know who those 'men of bronze' were, but we know with certainty that until the destruction that so transformed life on Crete, a busy sword-manufacturing workshop was exporting its products throughout the Aegean, the Dodecanese area and on the Greek mainland.[22] Those bronzesmiths did some of their finest work when it came to weapons technology.

Modern bronze uses other metals like zinc and manganese as an alloy. In very early times bronzes were made using arsenic. Arsenic bronzes have an advantage over pure copper in that they require a lower firing temperature, although they do not cast as well. But dealing with a deadly poison has its drawbacks: in the early days of bronze, smiths could not have lived very long to forge another day. Hephaestus, the god of technology and metalworking, held so important a place in the Greek pantheon of gods that he was generally portrayed as the son of Zeus and Hera – king and queen, as it were, of the gods. Yet he is also ugly and lame, a grotesque figure in many ways, and this folk memory could well reflect the effect that working with arsenical copper would have had on an individual.

When the French Emperor Napoleon died on St Helena in 1821, it turned out that he had been poisoned. But his death was almost certainly not an act of murder, despite the former dictator's numerous enemies. It was totally accidental. The mystery of why the Corsican died was finally solved when it was worked out that arsenic had been used to 'fix' the vivid green colour of the wallpaper of his genteel island prison. Arsenic can cause multi-organ failure and necrotic destruction of the body cells. Arsenic poisoning leads to central nervous system dysfunction and eventually death.

So the bronze that set the age of metals alight is a combination of 10 per cent tin and 90 per cent copper. Tin holds no danger for the metalsmith. That is by no means the only advantage of tin bronze. A sword forged with a combination of copper and tin, like those we had found in Heraklion, is strong but not brittle. Bronze is also malleable. It can be cast into a myriad of shapes and is therefore streets ahead of stone or wood as an easily worked and dependable material.

Curiously, the ores of both metals were readily available in Britain, but not in Crete. The strangest thing about bronze is that its two basic ingredients are rarely found in the same place. There is no source of tin in the Aegean. Yet the Minoans used the material with lavish abandon. Arthur Evans found massive tin bronze two-man saws at Knossos; huge ceremonial swords and axes were unearthed by peasants in the 1910s in the sacred cave of Arkalochori, and sold as scrap metal in the markets; vast cauldrons were discovered at Tylissos and Zakros. Yet either trade or travel was essential to make it. The advantage of creating a secure chain of ports to support voyages to find tin would have been obvious to any ruling élite.

As one expert, M. H. Wiener, says, bronze was so crucial that it would have been the object of intense search, planning and investment. He wrote that:

> The security, economy and hierarchy of Crete depended significantly on bronze. It seems inconceivable that ... Minoan palace rulers would have waited passively, hoping for a new Eastern Merchant-man to arrive with copper and tin.

Mighty Egypt, then, had no tin and only limited copper in ancient times – her mines only produced around 5 tons of copper a year. Egypt could not possibly have made the huge numbers of bronze saws required to build the pyramids unless either the raw materials or the completed bronze saws were imported.

So how did the Egyptians actually get the raw materials? Because this all predates widespread written records by many millennia, deciphering the historical record is a painstaking task. Lead isotope and archaeological evidence pins early sourcing of copper ore down to the island of Kythnos; Siphnos played the same role for supplies of silver and lead. Later in the Bronze Age, larger and more economical copper deposits on the Greek mainland and Cyprus were used. After doing some of my own digging and delving, I discovered that the Troodos mountains in Cyprus held ancient copper mines. It is generally thought that Cyprus was known at the time as the kingdom of Alashiya, which seems to have been a client state of the Hyksos (15th-Dynasty Egyptian rulers). It may also have been difficult to maintain a consistent supply of an ore which was in such high demand.

Copper was also found in Anatolia (modern-day Turkey) and Oman: at one point in prehistory there was also an important metals market on the island of Bahrain. Early Bronze Age tin workings have been found in Turkey's Taurus mountains, at Kultepe. However after 1784 BC there was no tin in the Mediterranean and not nearly enough copper.

Tin was the strategic prize. Generally the metal was only found far, far away from Crete – and there, I was sure, hung a tale. What if a sailing nation had taken the helm of the new bronze revolution? My research was all heading towards one conclusion: that the Bronze Age couldn't have happened without the existence of world trade. Who could have been behind that trade? The island of Thera was, in the Bronze Age, a very important place, the equal of Phaestos, Alexandria, Tell el Dab'a, Tyre or Sidon: not many ports would have had as many ships as can be seen on the admiral's frescoes.

Again, the evidence all pointed towards this one idea: to support the thriving trade, a maritime network had existed across the Aegean. The wonderful National Archaeological Museum in Athens holds a wealth of foreign artefacts found in tombs on the Dodecanese and Cyclades islands, notably on Thera and Samos, together with pictures of Bronze Age trading ships both painted and engraved on pottery.

That the Minoans had the ships, there was no doubt; but did their trading partners on the islands give them the scope to, in effect, rule the seas? How many Mediterranean Bronze Age ships were capable of following the Minoan ships to trade across the Mediterranean and beyond?

Historical sources – Homer's *Odyssey* and the later voyages of Pytheas the Greek – do suggest that maritime technology had been developing for many thousands of years before the Persians lost 300 ships in their famous first attempt to invade Greece (and 600 on their second attempt). Nearly 1,000 ships on both sides were involved in the second Persian assault. Although this was, of course, long after the Bronze Age, it does point to a long maritime tradition.

In order to try to make a reasoned estimate of the possible numbers of ships that had once plied the ancient seas, I started with the number of ports which existed in the Middle Bronze Age Mediterranean at around say, 1800–1500 BC, then I multiplied those ports by the number of ships each could have built. The Aegean has more than 1,400 islands and isles –'natural incubators for maritime technological development', according toWiener.

To give a couple of illustrations from these sources: Pytheas the Greek mentions Alexandria, Tyre, Sidon, Athens, Miletus, Apollonia, Odessus, Callatis, Olbia, Cumae, Nikala, Antipolis, Agde, Santa Pola and Carthage as established trading ports. We have Phoenician records for North African trade – at Ceuta, Melilla, Malaga, Algiers, Bizerte, Tunis, Tripoli and Sfax (to use later European names). The major Cretan palaces would probably each have had their own ships – Ayia Triadá, Phalasarum, Lisos, Souya, Prevéelli, Kommos, Lebèn, Myrtos, Iera Petra, Zakros, Sitei, Gournia, Malia, and Amnisos (Knossos' port). The Cypriot Bronze Age ports were Louni, Soli, Kyrenia, Peyia, Paphos, Corium, Limassol, Amathus, Larnaca, Agia Napia, Faralimni, Famagusta, Salamis and Trachinas. Adding these to the well-known Levantine and Black Sea ports suggests there were well in excess of seventy Mediterranean ports able to build and equip ships capable of sailing the Mediterranean and perhaps further (see maps).

How many ships would each port have built? This is almost impossible to answer. The Thera fresco shows ten ships, of which six appear to be ocean-going. For Thera to be able to have six ocean-going ships at sea at any one time, it is likely that she had at least another eighteen being built or repaired or under training – in short, a total of at least twenty-four Theran ships – let alone those docked at, or owned by, other Minoan ports.

As Homer describes Crete providing eighty ships from seven ports for the Trojan War, we may make a conservative estimate that each port built eight. There were seventy ports capable of shipbuilding. So we have around 560 – say 500 – ships capable of sailing the Mediterranean and beyond, able to carry the critical raw materials upon which the Bronze Age relied: copper and tin.

So the Minoans of ancient Crete had well-organised, well-planned cities; they had roads and ports; they had lighthouses; and they had ships. And those ships transported the goods they made – from honey to bronze tools, exquisite pottery and fine wine – which others wanted to buy. And on their fertile island heaven the Minoans happened to be very close to the rich markets of ancient Egypt and the Far East. Crete must have had fleets of ships sailing to Africa a full 4,000 years before Vasco da Gama and the 'Age of Discovery'.

Imagine a truly distant past. Crete is the magnificent crossroads linking three continents. On this strategically placed island the racial and cultural influences of Europe and Africa – and perhaps even Asia – all meet and mingle.

7

WHO WERE THE MINOANS?
THE DNA TRAIL

It was another newspaper report that prompted the next stage of my enquiry. 'Until now we only had the archaeological evidence [for Minoan genetic origins] – now we have genetic data too, and we can date the DNA.' I read this phrase on a Tuesday, over my morning coffee. By the Friday, I was on a plane to Istanbul.

Awoken at dawn by a deep-throated *muezzin*'s roar, I took a taxi to the hydrofoil in Istanbul's old Byzantine harbour. We skimmed across the Sea of Marmara (see map) to Yalova on Turkey's Asiatic Coast. At a spotless bus station with kiosks staffed by beautiful girls selling kebabs, tea and pastries, I discovered no fewer than 105 bays of buses, almost all of them huge and luxurious Mercedes vehicles. Grabbing one of them, I rode in air-conditioned comfort to the old Ottoman capital of Bursa, in western Anatolia.

There was at one time a theory that the Minoans' forebears were African. I knew that many experts disagreed: they believed that foreign settlers had arrived in Crete from southwest Anatolia, a people with a language related to that of the Hittites, further east.[23] There was a variety of evidence emerging from different fields of research. In 1961 Leonard Palmer noticed a link between Linear A and the Luvian language. In his book *Minoans*, Rodney Castleden argues that there may have been significant cultural contact between the Minoan and the Arzawa lands of southwestern Anatolia during the Second Temple period. The Hittite kingdoms at various points in history maintained strong contacts with Arzawa and Castleden mentions that the suffix '-me', which frequently crops up in Linear A associated with deity figures, in fact means 'lady' in the Luvian language. All of which would fit in with the Minoans' fervent goddess worship.

Now that scientists are able to test genetic theories with rigour, I was here because of the new study reported by *The Times*. New work by an international group of geneticists showed that a section of Crete's Neolithic population (i.e. pre-Bronze Age) did indeed go there by sea from Anatolia – modern-day Turkey. Professor Constantinos Triantafyllidis of Thessaloniki's Aristotle University had published the findings of a research group led by geneticists from Greece, the United States, Canada, Russia and Turkey. Professor Triantafyllidis said the analysis indicated that the arrival of these new peoples had coincided with a social and cultural upsurge that had led to the birth of the Minoan civilisation around 7000 BC. Specifically, the researchers connected the source population of ancient Crete to well-known Neolithic sites in Anatolia.

> The earliest Neolithic sites of Europe are located in Crete and main-land Greece. A debate persists concerning whether these farmers originated in neighbouring Anatolia and over the role of maritime colonisation. To address these issues 171 samples were collected from areas near three known early Neolithic settlement areas in Greece together with 193 samples from Crete. An analysis of Y-chromosome hectographs determined that the samples from the Greek Neolithic sites showed strong affinity to Balkan data, while Crete shows affinity with central/Mediterranean Anatolia. Haplo-group J2b–M12 was frequent in Thessaly and Greek Macedonia while haplogroup J2a–M410 was scarce. Alternatively, Crete, like Anatolia showed a high frequency of J2a-M410 and a low frequency of J2b-M12. This dichotomy parallels archaeobotanical evidence, specifically that white bread wheat (*Triticum aestivum*) is known from Neolithic Anatolia, Crete and southern Italy; [yet] it is absent from earliest Neolithic Greece.[24]

In the darkness before dawn we passed lines of pastry shops selling glacé fruit – oranges, black grapes, red cherries, auburn apricots. A charming city built on the slopes of Mount Olympus, Bursa is called

'The Emerald' with good reason. To her east lies the verdant and fertile Sakarya River valley, watered by streams from the mountains.

Forty-five minutes out of Bursa we began our ascent to the vast Anatolian plateau. The plain is lush and green with endless varieties of figs and a mass of apricots in the foothills. Flocks of black turkeys waddle beside the road. Here and there are plantations of poplars interspersed with oaks, pines and plane trees. Flocks of sheep and goats hold up the bus.

After two hours we reached the huge, rolling plateau of central Anatolia. Once the mass of apricots in the foothills had given way to poplars and plane trees, the trees disappeared to be replaced by long rolling wheat fields stretching to eternity, interspersed every 10 miles (15 kilometres) or so by giant sugar beet factories. Lines of greenery marching away to the distant mountains told of the track of watercourses. The houses in this area are huge – far bigger than in rural Greece or Romania to the north. Clearly Anatolia could feed tens of millions of people: it had so many natural advantages – rich soil, abundant water, plentiful sunshine.

For hour after hour we roared across this great plateau, heading for Boğazköy, 125 miles (200 kilometres) east of Ankara, which thousands of years ago had been the centre of the Hittite civilisation. I knew that name from the Bible: some of King David's most valued soldiers had been Hittites, although I'd never before troubled to find out exactly what that meant. In Boğazköy's museum I found seals with merchants' signatures, dating from 4000 BC. Arrows, axe heads and jewellery fill the rest of the museum.

Boğazköy is a higgledy-piggledy town with red pantile roofs and an antiquated *dolmus* bus that won't leave the square until it is totally full. This is where some of the very first Minoans may have come from. But then, you'd need a scorecard to keep track of all the peoples and cultures that have passed this way. The estimate is that the people whose arrival so transformed Neolithic Crete would have left this region around 100,000 BC.

The former Hittite capital of Hattuşa rises rather accusingly from the top of a huge crag, high above the present-day village. A hot walk up a tarmac road reveals an imposing lion gate but little else other than the rubble and footings of a ruined acropolis. The lions look strangely like puppies; obviously sculpted by someone who had never had the opportunity to study the beasts at first hand.

At its peak c.1344–1322 BC, this great palace was protected by 4 miles (6 kilometres) of stone walls. Now it looks about ready to turn back into dust.

To my right, as I walked through the ruins, was the outline of the oldest-known library ever to have existed. It is now nothing more than a few dents in the ground. I scrolled my toe through the dust and earth. Archaeologists tell us that all those years ago the cuneiform tablets stood on end in rows, like modern books on a shelf. There were even indexes, also marked out in the baked clay, telling you what was inside each stack of 'books': for instance, a label saying 'Thirty-two tablets concerning the Purulli festival of the city of Nerik'.

I wondered what it would have been like to go to a Purulli festival. Purulli was a storm god so I thought of the British music festivals, so often plagued by rain. Quite some way off from Glaston-bury, I suppose, but maybe not radically different – all that mud and all those brollies.

I was glad, then, that I'd gulped down my coffee and headed for the airport. A whole new way of investigating the Minoans' past had now opened up to me: through DNA.

The Hittites must have been a formidable people. They signed peace treaties with Egyptian pharaohs and Assyrian kings. They also conquered Babylon and were a well-ordered society, worshipping many deities, including, I noticed, a female god – Hepatu, the sun goddess. I didn't know for certain whether that was a direct link to the powerful goddesses of Knossos, but it felt right. Just to the south of here, on the Konya plain, is Çatalhöyük, known as the world's first major human settlement, of c. 7500 BC to 5700 BC.

Turkey not only straddles two continents, it seems. It straddles time itself.

|||||||||||||||||||||

What was the link to ancient Crete and to the Minoan civilisation that once flourished there? Very early settlers to Crete introduced cattle, sheep, goats, pigs and dogs, as well as vegetables and cereals. The earliest communities grew up around the coast on the eastern and southern parts of the island. Yet something appears to have happened to Crete and its culture after 7000 BC – something momentous. People suddenly learned advanced practices in agriculture. They developed pot-making, and metalwork. Next, quite suddenly, came an incredible period of palace development. How did such amazing sophistication appear so suddenly?

Boning up on the research I'd first read about in the newspaper report, I discovered that the DNA haplogroups J2a1h-M319 (8.8 per cent) and J2a1b1-M92 (2.6 per cent) linked the Minoans to a Late Neolithic/Early Bronze Age migration to Crete no later than 100,000 BC. Specifically, genetic researchers connected the source population of ancient Crete to the well-known Neolithic sites of ancient Anatolia, not too far away from my present location – such as Asýklý Höyük, Çatalhöyük, and Hacýlar. That made the blood surge a little faster. If researchers could find and track an ancient DNA link to Crete from the region I was now visiting, then perhaps I could follow the Minoans – via their DNA – to some of the other places I suspected they'd been. And in so doing reinforce my notion that they'd been the most important international trading culture of their age. And as such a key defining force in global history.

All human beings carry DNA (deoxyribonucleic acid) in every single cell of their bodies. Each cell contains 46 chromosomes, which cluster in pairs: half derive from your mother and half from your father. Chromosomes contain tightly coiled DNA, divided into sections known as genes. Genes tell the cells what proteins to make. The proteins, in turn, control everything that happens in your body when it comes to identity and growth: for example, one protein might make your eye pigment, others might decide the size and shape of your teeth.

The paired DNA strands wrap around one another to make a double helix. Each person has their own unique genetic fingerprint. It is unlike the DNA of anyone else, but there is a useful rider to that. While women have two X chromosomes, men have one X and one Y. The Y chromosome is always passed down the male line from father to son. That unique factor makes the Y-line virtually like a surname: it is transmitted, almost intact, down the male line from generation to generation.

The Y chromosome is altered only by rare spontaneous mutations. These mutations can be used to identify sequences of the Y chromosome, known as haplogroups. Being no expert in DNA, I had to do a bit of reading up here. The word 'haplogroup' comes from the Greek and means single, or simple. Haplogroups reveal deep ancestral origins dating back thousands of years. Males with the same haplogroup must have shared a common male ancestor in the past. This was intriguing, because according to what I was now learning other great peoples, such as the Etruscans, seemed to share some common ancestry with the Hittites of Anatolia. Haplogroups also incorporate smaller Y-DNA sequences know as haplotypes – groups of genes that share a common ancestor. Thanks to the work of the Human Genome Project, many of these have been identified and given codenames, e.g. J2a-M410.

It was Professor Constantinos Triantafyllidis of Thessaloniki's Aristotle University who had released details of the research and it was his comment that had launched me from my lukewarm morning coffee to a blasting hot plain in central Turkey. According to his initial findings, today's Cretan population was genetically intermingled with yesterday's people of Anatolia. Professor Triantafyllidis believes the arrival of these people had coincided with a social and cultural upsurge. It led, around 7000 BC, to the birth of Europe's first advanced civilisation.

I remembered very clearly one surprising image imprinted on the Phaestos Disc. It was of a man wearing a striking feather headdress. My trip to Turkey had confirmed my suspicion: during the Bronze Age, such warlike headgear was worn by the Lycians of Anatolia.

Notes to Book I

1. Marthari personal communication; S. Marinatos 1974, 31 and Pl. 67b and d; C. Doumas. C., in *Thera and the Ancient World*, 1983, p. 43

2. Rodney Castleden, *Minoans: Life in Bronze Age Crete*, Taylor and Francis, 2007

3. M. H. Wiener, *Thera and the Aegean World III*, vol. 1, *Archaeology*, Proceedings of the Third International Congress, Santorini, Greece, 3–9 September 1989, p. 128

4. Ibid

5. Papapostolou, L, Godart and J. P. Olivier, pp. 146–7, *Roundels among Minoan seals* (2009)

6. Plato, *Timaeus*, trans. Robin Waterfield. Oxford World Classics, 2008

7. Ibid

8. J. Boardman & colleagues, 'The Olive in the Mediterranean: Its Culture and Use', Royal Society Publishing, vol. 275, No.936, JSTOR

9. Ibid

10. Ventris and Chadwick, *The Decipherment of Linear B*, Cambridge University Press, 1958

11. Shaw, B. D., *The Cambridge Ancient History*, Cambridge University Press, 1984

12. Stylianos, Alexiou, *Minoan Civilisation*, Spyros Alexiou Sons; First Edition (1969)

13. Bruins, MacGillivray, Synolakis, Benjamini, Keller, Kisch, Klugel, and van der Plicht, 'Geoarchaeological tsunami deposits at Palaikastro (Crete) and the Late Minoan IA eruption of Santorini', Journal of Archaeological Science 2008, 35, pp. 191–212

14. Joseph Shaw (www.fineart.utoronto.ca/kommos/kommosintroduction)

15. Thucydides 1.41. trans. Benjamin Jowett. Oxford, 1900

16. Bernard Knapp, 'Thalassocracies in Bronze Age Eastern Mediterranean Trade: making and breaking a myth', in *World Archaeology*, vol. 24, No.3, Ancient Trade: New Perspectives, 1993

17. Thucydides 1.9.1,3

18. Stylianos Alexiou, *Minoan Civilization*, trans. C. Ridley. Heraklion Museum, 1969

19. Robert Graves, *The Greek Myths: Complete Edition*, Penguin, 1993

20. Stylianos Alexiou, *Minoan Civilization*

21. Herodotus, *Histories*. trans. George Rawlinson, Penguin Classics, 1858

22. Sandars 1963, p. 117; Popham et al. 1974, p. 252; Driessen and Macdonald 1984, pp. 49–74, 152 in "*The Isles of Crete? The Minoan Thalassocracy Revisited*, The Thera Foundation (www.therafoundation.org)

23. Sinclair Hood, in *Archaeology: the Minoans of America*, vol. 74, 1972

24. Constantinos Triantafyllidis, Aristotle University, Thessaloniki

BOOK II

||||||||||||||||||||||||||||||||

EXPLORATION

VOYAGES TO THE
NEAR EAST

8

THE LOST WRECK AND THE
BURIED TREASURE TROVE

By now I was becoming convinced the Minoans were the forebears of the mythic heroes who sailed to Troy: Agamemnon, Achilles, Odysseus et al. But without any actual physical evidence of the kinds of ships they would have sailed, I feared that the trail had run out. Then one night I heard about a wreck dating from c.1305 BC that had been found on the seabed of the nearby Turkish coast – a discovery which, to a former sailor like me, counts as one of the greatest archaeological finds ever made.

In sea terms Uluburun, a promontory off the coast south of Bodrum, is quite close to Thera. Significantly for me, it is in exactly the zone – western Anatolia – that Bernard Knapp had described as being under the control of the Minoan empire. In Turkish the term 'Uluburun' simply means a rocky plinth of land; but the reef below it, like all reefs, is treacherous in bad weather.

I arrived at nearby Bodrum early in the morning, having spent an uncomfortable night chugging along the Aegean on yet another ferry. To get here, we puffed in the opposite direction from the beautiful Sea of Marmara and the teeming, anarchic sights of Istanbul. A litter of rocky islands stood in our way; as we puttered forward, I thought you'd have had to be a pretty good navigator just to travel north–south across the Aegean Sea.

We awoke to a fresh wind and the unmistakable bustle of a boat nearing port. I rushed out on deck in the sharp air of the early morning to catch my first glimpse of coastal Turkey for more than twenty years. Purple hills, deep grey-green olive groves, the azure sea and the romance of the deep past: I breathed it all in.

Excitement made my skin prickle. I could be on the threshold of a real breakthrough in my quest to uncover the truth behind the explorers and the seafarers of the ancient world – today at the town museum I'd get my first view of the oldest-known shipwreck in the world.

Bodrum is more or less the home of history; at least, it was home to the world's first superstar historian, Herodotus. The town's ancient name was Halicarnassus and its walls once encircled a mausoleum that was one of the famed seven wonders of the ancient world. It was such a pleasure to be here and breathe it all in; you can still trace the ancient walls that once embraced nearly the entire town. Bodrum Castle, now the Museum of Underwater Archae-ology, was built by the Knights of St John in the 15th century. I felt instantly connected to the castle: its five major towers include the English, or Lion, Tower. Colourful rows of huge amphorae are mounted on the entrance wall.

The Uluburun wreck is the kind of heart-stopping find that archaeologists dream about; one of those discoveries that helps rewrite history. Quite why the boat came to grief – near what is today the pretty little fishing village of Kaş – no one will ever know for sure. The ship foundered a hundred years or so after Thera's frescoes were painted. Yet this is hardly a blink in time: in the interim, shipbuilding technology would not have changed much at all.

Poignantly, the wreck held one of ancient history's most extra-ordinary finds: the earliest 'captain's logbook'. But the captain remains mute, his words – written in soft wax – obliterated by time.

||||||||||||||||||||

For centuries the wreck lay quiet, in total obscurity, protected from looters by 45 deep blue metres (150 feet) of silent sea. Then in 1982, thousands of years after its last journey, a sponge diver found it by total accident. We must thank the stars for Mehmet Cakir's decency and honesty. If it were not for him, some of the world's greatest treasures could have gone via the black market to unscrupulous

74

dealers. For what Cakir found was a hoard worthy of an Aladdin's cave – mound upon mound of sunken treasure.

The ship's extraordinary cargo could have come from a modern French luxury goods house – except that these opulent objects were much more exotic than even the highest quality champagne. The treasure, found scattered over 250 square metres (300 square yards) of the jagged, rocky seabed, came from as many as ten different Bronze Age nations. The Uluburun ship carried exquisitely wrought gold and silver jewellery and a cornucopia of rich fruits and spices, as well as huge amphorae from the Lebanon; terebinth resin, used to create perfumes; ebony, which had come all the way from Egypt; and all manner of exotic goods like elephant tusks, hippopotamus teeth and ostrich and tortoise shells (see first colour plate section).

Before you see the ship, the museum guides you through the finds from its hold. The material found in shipwrecks can only be rescued with a huge amount of hard work. Firstly, artefacts are put in vats of water, where they will remain for approximately six to eight years. It takes a minimum of five years, and sometimes up to ten years, just to remove the salt from porous substances. The water is changed every fifteen days, depending on what the gauges indicate. After the salt is removed, the objects are placed in a vat containing polyglycol for a period of four to five years. Every possible care has been taken to save these precious things. It was not difficult to see why.

The greatest treasure, a prize find worthy of a king, lay at the stern. It was an extraordinary chalice of pure gold. Homer describes the legendary war hero Achilles drinking from a golden goblet just like it:

> Achilles strode back to his shelter now
> and opened the lid of the princely inlaid sea chest
> that glistening-footed Thetis stowed in his ship to carry,
> filled to the brim with war-shirts, windproof cloaks
> and heavy fleecy rugs. And there it rested . . .
> his handsome, well-wrought cup.[1]

Achilles' golden chalice was found nestling on the seabed next to an exotic pendant in the form of a golden falcon, which was clutching a giant cobra in its talons. In the hold was yet more treasure, including a gold scarab bearing the name of the celebrated Egyptian beauty Nefertiti, queen and wife to Akhenaten, the all-powerful pharaoh of Egypt.

But in many ways I was less interested in all this exotica than in the 'everyday' story this traveller could tell us. The ship was a time capsule, speeding us back to the daily life of the Bronze Age: for me, it would be almost as if I were stepping into Doctor Who's Tardis.

As I travelled along the solemn line of glass cases, I could see knives and weighing scales – practical tools that had been used by ordinary sailors thousands of years ago. There were many animal-shaped lumps of bronze: deer, cattle. They stood out because they looked like toys, but in the catalogue they were described as weights. The weights used for scales were particularly eye-catching, even charming. There were stone mortars for grinding food and fish hooks to catch it. There was a barbed trident: it made you think of the Roman *retiarii*, or net and trident fighters – introduced to a much later Roman empire by Emperor Augustus.

The *retiarius* was modelled on a fisherman: the edges of the net he used like a lasso were weighted with lead. Right in front of me were the Bronze Age predecessors of that military innovation; hundreds of net sinkers, the weights that the crew used on the edges of their casting nets. There was also a harpoon – and a set of knucklebones, for playing a game of chance on those long nights at sea, a night's entertainment lit by the comforting yellow glow of oil lamps. Weapons, too – arms that had been wielded by wizened and ancient hands more than 3,000 years ago.

Then I reached the last display case. Its contents, for me, were far more valuable than all of the silver and gold, however beautiful. I took a sudden deep breath.

I was right. What was the wreck's main cargo? The raw materials for making bronze. They were even in the right proportions: 10 tons of copper and 1 ton of tin.

This was truly a Bronze Age vessel, made with bronze tools and carrying enough copper and tin to make weapons for an army. There were 121 copper 'bun' ingots, together with enough fragments to make another 9, and 354 copper 'oxhide' ingots, with an average weight of 23 kilos. Each copper oxhide ingot was about a metre long, with sharp corners. They are called oxhide ingots because their distinctive shape is a bit like the hide of an ox when it has been flayed and stretched. At one time it was thought that they may have represented the value of an ox; now archaeologists think the shape is simply easy to handle.

ıιιιιιιιιιιιιιιιιιιιι

I slowed down as I approached the next room. Ahead of me was what I had been looking for: part of the oldest ship I had ever seen. About 15 metres (50 feet) long, her charred timbers were spindly and frail. She may have been ravaged by centuries of seawater, but there was a strength and grace to this boat's design that made you forget about her current dishevelled state. Here was a ship that could sail like the wind; nimble, responsive and manoeuvrable. There were quite a few other visitors in the room, but the atmosphere was hushed and thoughtful: after all, on this vessel people had struggled for their lives. And lost.

It took an astonishing eleven years and 22,000 dives to excavate the Uluburun. The ship had come to rest listing 15 degrees to starboard, facing east–west. A large piece of the hull had lain intact on the seabed for almost 3,500 years. It seemed incredible to me that it had survived at all.

The ship was built of cedar in the ancient hull-first tradition, with pegged tenon joints securing planks to each other and then to the keel. To build her, a team of men would have cut down a tall cypress tree, stripped its bark and carved the length into a keel.[2] Crete was covered with cypress forests in the Early Bronze Age: the wood is ideal for shipbuilding because it expands in water, making the seals between joints waterproof.

The shipwrights then 'edge-joined' a long cleanly sawn plank of cypress to each side of the huge keel. Chiselling out deep rectangular matching mortises every 25 centimetres (10 inches), they worked along the length of the keel. They cut flat rectangular pieces of wood (tenons) to fit snugly into the matching mortise slots. Then they fitted another plank to the tenons that were sticking out of the slots on the keel.

Methodically, the shipwrights added plank after plank to both sides of the keel, to build up a sturdy hull. The boat needed very little caulking given the edge-joined planking, although they did waterproof the seams with a resin mixture. The shipwright who worked for the mythological hero Jason, Argus, had cut his timber on Mount Pelion: but according to Thucydides, the search for wood for battle fleets meant that the plentiful forests of Greece were stripped bare as early as the 5th century BC. The Uluburun's sturdy oak mast was about 16 metres (52 feet) tall and secured with rigging made of strong hemp ropes. As someone who had spent many years at sea, I was sure that boats like these were fully capable of surviving long and hazardous ocean voyages, despite this individual vessel's fate.

But could I prove it? Were the exploits of ancient adventurers simply the stuff of legend, or could they be close to the facts? Luckily, I had an ancient myth and a modern real-life explorer's vessel to compare it with. Scholars say the legendary 1,500-mile voyage of Jason and the Argonauts, to win the Golden Fleece, happened during the Bronze Age, if it ever really took place. In the 1980s Tim Severin became obsessed with recreating a boat that could test the reality behind the myth.

The myth itself is a sort of ancient Greek *Mission Impossible*. The hero's kingdom is under threat from a usurper, his uncle, King Pelias. The gods command Jason to find a magical ram's fleece before he can reclaim his father's kingdom of Iolkos. So Jason has to embark on a treacherous sea voyage into an unknown land.

To test whether the tale could ever have held any truth, Severin reconstructed an ancient galley with documentary precision. With oars as well as sails, a galley is the logical choice for a vessel that might have to fight its way through hostile waters. I seemed to remember that Severin had managed it as far as the Black Sea without sinking – although there had been some pretty tricky moments.

The Uluburun ship would have fared even better. She was an altogether bigger and more seaworthy affair. The ship's cargo alone told us that her owners felt she could travel far and fast. It included Baltic amber from the north; ebony wood, hippopotamus teeth and elephant tusks from equatorial Africa; and goods from both the eastern and the western Mediterranean. The voyage itself almost certainly included stops in Cyprus for copper, Egypt for gold and Syria for ostrich eggs and pomegranates. That involved steering to all parts of the compass. If Tim Severin could make it, then so could the ship I had in front of me in Bodrum. The Uluburun wreck may have gone down like the *Titanic*, but that was because the captain was negotiating a dangerous promontory, perhaps braving the greater dangers of sailing in winter. In normal conditions, she was definitely able to weather even high sea states.

Archaeologists agree that the discovery of the Uluburun has pushed back the history of seafaring by centuries. I say that the evidence of the frescoes on Thera pushes that history back much, much further. Given the dating of the frescoes, which show ships of the exact same construction, there must have been a long tradition of sailing – and the know-how was alive and in use by the time the artists put paint to plaster. Shipbuilding technology changes, but it does so only slowly – or did, until the breakthroughs that came with the Industrial Revolution: steel and steam. Had he been alive, the expert mariner Sir Francis Drake – whose famous *Golden Hind* was a round-bellied Tudor ship – would still have been able to make a good fist of sailing Admiral Lord Nelson's much larger and slimmer HMS *Victory* at Trafalgar in 1805, some two hundred years later.

The Uluburun's mast had a single boom about 10 metres (33

feet) long to hold the top of the sail. The centre of the boom was connected to the mast by a thick strong ring of rope, loosely wrapped around the shaft of the mast (see also Thera ship 6). This meant the boom could pivot freely about the mast in the wind. It could be easily raised or lowered with ropes running through a bronze fixture on the masthead. The orientation of the boom and the sail in the wind could be controlled from the deck with ropes, allowing the ship to tack quite well into a head wind.

Thinking back to the frescoes, the evidence here in Bodrum confirmed what I had already noticed. Each of the long ships was a masterpiece: I was overwhelmed. All of the evidence told me that shipwrights were definitely capable of building ships fit for ocean travel as early as c.1450 BC. The sailors had sophisticated methods of hoisting, lowering and adjusting sails, not least the bronze fixture on the masthead. There is other evidence of this impressive development in technology: the mechanical principle is shown by two separate examples in the Egyptian Gallery in the National Archaeological Museum in Athens. Those fixings show that many centuries ago ships were confidently expected to sail into the wind as well as before it: the men knew they could lower sail very quickly in the event of an unexpected squall. By comparison with the Uluburun's hold, I estimated that the cargo capacity of the admiral's ship at Thera would have been about 50 metric tons.

Using bronze tools, particularly the large two-handled saws that were almost 2 metres (6 feet 6 inches) long and a third of a metre (13 inches) wide, these ancient craftsmen had achieved levels of shipbuilding expertise that Renaissance Europe had taken centuries to rediscover. This was a Bronze Age civilisation. But making the most of Bronze Age technology, it seemed, was not exactly making the best of a bad job.

The museum held another surprise. The Minoans have always been characterised by archaeologists as a wholly peaceful people. If so, their level of weaponry is highly surprising. Arthur Evans had bitter experience of violence, witnessing massacres in Crete as the *Manchester Guardian*'s war correspondent during the 1897 Greco-

Turkish war. After discovering Knossos, Evans declared that by contrast his beloved Minoan civilisation had been totally peaceable. As Cambridge University research scholar Cathy Gere describes it:

> Evans had reported with fairly even-handed revulsion the terrible Muslim/Christian massacres on Crete. Keen to represent the past as a site of healing and reconciliation, Evans resurrected Bronze Age Crete as an unfortified idyll, internally peaceful under the benign reign of Knossos, and protected from its enemies without by the legendary seafaring skills of King Minos' navy.[3]

Since then, most academics have followed his lead and depicted Crete as a peaceful society, mainly because its cities didn't have fortified walls. The Minoans lived a harmonious life in lush, civilised surroundings, they say, calmly governed in a democratic way. They didn't need city walls. Their kingdoms resembled peaceful heavens on earth.

Yet as I moved through the closeted dark of the exhibition rooms, I came upon case after case of weapons recovered from the wreck. Bronze weapons, manufactured in large quantities. Green and corroded as they were from centuries on the deep seabed, I could clearly see that this prehistoric ship had carried everything a troop of warriors would need – arrowheads, double-headed axes, spears, swords and daggers as well as the adzes, saw screws and razors that had more peaceable uses.

Crete's largest archaeological museum, at Heraklion, also has cases full of ordinary soldiers' bronze weapons – from spears to daggers and nasty-looking rapiers. Some of them are obviously ceremonial – like the extraordinary dagger found at the town of Malia, its hilt decorated with gold leaf; but towards the end of the period it was plain that swords had been made for only one purpose: fighting. Short blades had been adapted for cutting, as well as thrusting strokes. They were made for warriors.

Without doubt, the Minoans' military strength came from their fleet. Coupled with the sophisticated design of their ships, it seemed to me that far from being entirely peaceable, the Minoans had possessed formidable military might. The warlike Mycenaeans on the mainland of Greece would later wage war against the Trojans, as described in the *Iliad* and the *Odyssey*. Yet the Minoans seemed to live at peace with them. In fact, much of the archaeological record points to the idea that the mainlanders were either a Minoan colony or a compliant ally of the Minoans – until the point at which Crete was fatally weakened by the devastating loss of its wooden fleet in the tsunami.

Of course the Minoans may have preferred not to use force: who knows? They were certainly as interested in trade as they were in weapons. Yet it was noticeable that their fast, sturdy ships were the type that could be mobilised speedily, carrying soldiers armed to the teeth with a lethal array of bronze weapons of war.

Like most British citizens of my age, whose parents had been through a war, I understood the strategic value of being 'an island nation'. The sailors of Atlantis were already the world's top naval power. With ships that were far in advance of any others in Europe or the Near East, the result must have been that the Minoans feared no one. Their remarkable cities had few defensive walls precisely because they had the sea instead. Leaving them wide open to the biggest tidal wave the world had ever seen.

At least one of Thera's marvellous frescoes shows soldiers in battle helmets getting ready for war, along with an admiral commanding a sea battle. And winning it. They'd evidently had the might to control the entire Mediterranean. Maybe they'd used it.

And that's what I was determined to find out next.

||||||||||||||||||||||

In construction, the Uluburun wreck had been almost identical to ship 5 on the fresco in the West House. That meant one thing was certain: the Minoans had similar, deep-sea trading ships – at least a hundred years before the ship that had gone down off the coast of Turkey. I left the calm and quiet gloom of the museum with the blood spinning in my head. Ostrich eggs from Syria; gold from Egypt; precious amber from the Baltic; hippopotamus teeth from equatorial Africa ... all of the riches of the world. All here.

Here in Bodrum I felt I was looking at positive proof that the Minoans, these highly creative, ingenious and artistic people, were also the world's first explorers.

9

SAILING FROM BYZANTIUM

I'd left Turkey and the Uluburun wreck with a sense of sadness. The captain's book, that simple writing tablet in folding wood, dominated my thoughts. The captain had loaded all of the riches of the world on to that boat – from tortoise shells for making musical instruments, and the much prized blue cobalt glass, to the smelly murex shells that were collected in their millions on Crete, and which produce the world's most brilliant purple dye. But nothing could protect either him or his passengers and crew from the perils of the sea. Even those with the most extraordinary sailing skills must wrestle with the elements, and rocky reefs, from time to time. And sometimes they will lose the fight.

The world's oldest book was made of two boxwood leaves, joined by a three-piece cylindrical hinge in ivory. The cavity was inlaid with wax. But again, like the Phaestos Disc, the words themselves escaped me. The captain was mute, his writing scoured away by the seas that had killed him. I could tell nothing whatever about his journey, or the skills he had as a sailor. The very fact that he had set out, though, was proof paramount that Bronze Age sailors had considered themselves capable of ambitious and wide-ranging journeys at sea.

The book must have held instructions from the ship's owner, or even from the ruler, I thought, about the cargo and where it was headed. It would read something like this: 'The captain is to proceed to Byblos from Thera and unload the cargo. This special cargo is the property of X, and consists of 340 ingots of copper, and 60 of tin . . .'. Then would follow a detailed list, written with a bronze stylus,

84

of all the valuable articles on board, as well as weights, measures and destinations. I fell to thinking: if a thief had wanted to steal that captain's cargo, or if the captain himself had been dishonest, then wax is not exactly the ideal recording medium. A heated spatula would easily smooth it out, making the figures easy to alter.

The log – for that's what it must have been – was tied closed. It would also have been sealed with the owner's personal seal. And so that the system was foolproof, the owner and the captain used a common system of weights. On that fateful day the weights and at least three sets of balances were on the ship, suggesting that there were a number of different merchants taking passage on the journey. Already, in the Bronze Age, a single mass standard was used across the Mediterranean, based on a unit weight of 9.3 to 9.4 grams.

Sceptics may think that, surely, this skill with sailing could not have come suddenly, from nowhere, as if by magic. True enough: long-distance sailing is demanding. It takes great skill, intimate knowledge of the seas and constant vigilance, not forgetting bravery, perseverance and a certain amount of luck. When Tim Severin set out on the trail of the Golden Fleece, pessimists calculated that unless they had favourable winds to help them, the crew would have to row more than a million oar strokes each to reach Georgia. A testing mission.

This scepticism would be misplaced. When we 'discover' a new city, a new monument, or a new wreck, we don't manage to do it in neat, chronological order. To early archaeologists and explorers the ancient Egyptians must have seemed to have had almost pre-ternatural abilities. To many of us, they still do. Yet their technology wasn't sent down to earth by aliens, as some have claimed. The truth is that those achievements look surprisingly sudden to us because we are missing the various stages of development in between. In the case of the Egyptians, many sites dating from before Old Kingdom Egypt simply hadn't been dug up – or recognised for what they really were – until relatively recently.

Just like the hidden passages and chambers in the great Valley of the Kings tombs of the Egyptians, the Minoans' seafaring skills didn't come from thin air. This resourceful culture had built up a body of knowledge over the centuries, as I'd found when tracing their DNA trail back to the Hittites. Just as in Egypt, the longer we wait, the more evidence of early Cretan seafaring we will find.

I had to know more about Minoan sail-power.

Ships were essential to move the raw materials on which Minoan civilisation was based. They had to shift their finished works for export and they also needed to transport the raw materials for their products: whether that meant copper from Cyprus, elephant and hippopotamus ivory from India and Africa, gold from Upper Egypt, gypsum and glass from the Levant or the mysterious substance amber from the Baltic. They therefore needed ships that could travel hundreds of miles across the Mediterranean and, in the case of the precious amber, to the Baltic. Could they possibly have done that?

So I turned my thoughts back towards the actual task of sailing and to Tim Severin.

VOYAGE TO THE BLACK SEA BY REPLICA SHIP

A Harkness Fellow at the universities of California, Minnesota and Harvard, Tim Severin has written many books on exploration, from *Tracking Marco Polo* to *Vanishing Primitive Man* and, most importantly for me, *The Jason Voyage*.[4]

When Homer wrote the *Odyssey*, around the 8th century BC, he said that Jason and the Argonauts' brave and epic voyage in search of the Golden Fleece was already 'a tale on all men's lips'. This would have been a generation before the Trojan War. Critics have been sceptical. The voyage couldn't have had any basis in reality, they argued, because this was too far back in time; the technology simply wasn't in existence. Severin's aim had been to find out if Jason could have really sailed to the Baltic in 1300 BC. To quote him:

> ... we hoped to track them [the Argonauts] in reality. So we rowed aboard the replica of a galley of Jason's day, a twenty oared vessel of 3,000 year old design, in order to seek our own golden fleece ...

He based his reconstructed ship upon drawings on Bronze Age pots and frescoes, as well as carvings on armour, jewellery and seals. For reasons of cost it was a half-scale model, 16.5 metres (54 feet) long 'from the tip of her curious snout-like ram in the bows to the graceful sweep of her tail'. Later pictures of ships dating from the 7th to the 4th centuries BC gave Severin further technical details to follow, such as the way the sails were rigged and controlled.

In Homer's writings, Greek ships are measured by the number of oars. They came in three sets of twenty, thirty and fifty oars. To save money, Severin chose the smallest.

Severin was going against the currents all the way, with no help from the prevailing wind. The ship he would sail would be some two centuries younger than those shown on Thera's frescoes. It was a difficult task: although his ship was just half the length and breadth of ships 5 and 8 at Thera, with half the number of oars, its volume would be just one eighth of that of the Thera ships. This placed a considerable strain on cargo capacity, water and food – and on the crew. Because of the lack of space, sleeping was hard and the crew were permanently exhausted as a result.

Along with naval architect Colin Mudie, Severin examined every aspect of ships built in the area more than 3,000 years ago and how to sail them – not least in relation to the crew. The physique of man in the Bronze Age was smaller than today. Cretans were also small – they were perhaps 165 centimetres (5 feet 5 inches) tall and about a stone lighter than today. The pair searched for a boatbuilder who specialised in building traditional Aegean sailing ships, one who would work in the same wood that was used thousands of years ago. Vasilis Delimitros was reputed to be the best shipwright on Spetses, which is now a picturesque, car-free island and a splendid holiday haven. In Jason's time it would have been a Mycenaean stronghold. The galley was built using the same bronze adzes and saws which were in use in the Bronze Age. Delimitros followed Colin Mudie's drawings closely, not least when constructing the hull, where the planks were joined edge to edge as in the Uluburun wreck. Grooves were chiselled out of the plank, a tongue of oak was inserted, a matching tongue was cut out of the plank to be joined, the two were hammered flush together, then a hole was drilled for a locking peg through the matching tongues. To Vasilis it was no problem. He built one complete side of the ship by eye, then the next. The planks were slotted together as quickly and smoothly as if he had been doing the job all his life.

When the hull was ready they carved the mast from a tall, straight cypress tree, as well as two 3.5-metre (12-foot) high steering oars. The rigging was similar to ship 6 in the Thera fresco – hemp woven and stitched to make stays, sheets, and halyards. The pulleys were copied from the oldest-known shipwrecks, with wooden sheaves fixed around wooden pins. Finally the fierce and brave *Argo* was ready – and was even painted in the livery of red, white and blue found both in Mycenaean ships and on the Theran wall frescoes.

What now follow are my own impressions. At times they may appear critical: that is not the intention. I have unqualified admiration for what Severin achieved with a largely amateur crew on a journey of 1,500 miles (2,400 kilometres) across open sea. The personal toll was extreme. For instance, Severin describes an occasion when a gnawing storm brewed up in the Black Sea, and the heavy downpour turned the hands of the rowers a horrible, pulpy white. As he describes it: 'Their hard-won blisters looked like dead flesh.'

The incident itself clearly shows the dangers that Bronze Age sailors would have had to face routinely. Severin was attempting to reach a calm cove, against the pressure of great round-shouldered waves, through a narrow cut in the cliffs:

> We were getting dangerously near the rocks, and the crew were falling quieter as the constant slog began to sap their energy . . . I turned *Argo* at right-angles to the entrance and drove her hard at the gap. She rose on the back of a wave, heaved forward, fell back and was picked up a second time. The crew rowed flat out to keep her moving through the water so that I had enough speed to steer her and to keep the boat from wavering off-course or broaching to

the waves. In a spectacular roller-coaster ride the galley went tearing through the gap, her fierce eyes staring straight ahead, and we entered the haven.[5]

There were plenty of disasters, but in the event many things went better than planned. Firstly, the bow ram proved a convenient place for the crew to sit to relieve themselves. A line of pegs doubled as a ladder to climb back on board. The sailors could wash themselves and their clothes when becalmed – as well as fish from the ram.

Secondly, the twin steering oars (which were worked in opposite directions), turned the ship easily. Then in flat calm seas without wind, twelve oarsmen (out of the twenty) could push the ship along at 3 to 3.5 knots, even faster if the centre of gravity was adjusted by shifting the resting oarsmen and cargo aft. Severin writes:

> *Argo* spent all that day skimming over the waves in Jason's wake like a ship out of a dream. Under full sail she behaved superbly. The hull cleaved through the blue sea and with the merest touch on the two steering oars, *Argo* responded like a well-schooled thoroughbred. She turned as deftly as one could wish, and a second touch on the tillers to bring them level brought her back on course, running sweetly forward. One could feel the ship quiver as she sped downwind, her keel and planks thrumming as she rolled over to the following seas while her ram threw aside a cresting bow wave … For hour after hour *Argo* ploughed along at 5–6 knots leaving a clean wake behind her, while the crew relaxed on the benches …

Finally, *Argo* had very good sea-keeping qualities in a gale. Although she was wallowing and lurching, and was seemingly threatened by wave after wave, she did not take on much water. The crew – although they got seasick – just sat and waited for the gale to pass. *Argo*'s natural position was to lie broadside on to the waves, rising and falling to their advance and giving a lurching sideways heave as each crest passed under her keel.

Now for the bad news. Rowing had been a nightmare. Continuously soaked, chafed and wet, the crew suffered multiple salt-water sores – their hands a mass of blisters. Even when they'd replaced mutton fat with olive oil as a lubricant, rowing was only practical in a flat calm. The slightest wind was a nightmare – ' . . . a scarcely perceptible breeze blowing against the prow of the boat cut down her speed alarmingly. It was not just like walking uphill, but like walking uphill through shifting sand'.

In any sort of sea the oars got banged about and rowers were thumped in the back. The steering oars were not strong enough and broke frequently. Neither were the rowlocks, which snapped. The cotton sail became mildewed with damp (I think using a cotton sail was a mistake: Minoans used woven sheep's wool impregnated with fat).

After only five months' sailing the hull of the ship was covered in barnacles, reducing speed by one knot (the Thera frescoes show that the Minoans sheathed the hull in canvas, which would have eliminated most barnacles). Despite all of this the *Argo* covered 1,500 miles (2,400 kilometres) in three months – against the wind and current – with a constantly changing, inexperienced crew. It was an astonishing achievement.

IIIIIIIIIIIIIIIIIII

Severin and his crew had shown that boats like the *Argo* could have successfully plied the Mediterranean in the Bronze Age. By this evidence, they could even have got as far as the Baltic. Yet there were still major questions that needed answering, as far as I was concerned. I particularly wanted to discover whether it was specifically the Minoans who had developed the art of seafaring? Did they have a far-reaching tradition that made them better sailors than, say, the peoples from mainland Greece?

In spring 2010, I once again picked up my *Times* newspaper, and found, by luck, another piece of the puzzle.

Quartz tools at least 130,000 years old, such as hand axes, cleavers and scrapers, had just been found in an area of southwest Crete from Plakias to Ayios Pavlos, including the famous Preveli Gorge. They dated from the Lower Palaeolithic period. The fact that Crete has been isolated from the mainland by the Mediterranean Sea for five million years told the archaeological team that these early settlers must have arrived by boat. The team, headed by Dr Thomas Strasser of Providence College Rhode Island and Dr Eleni Panagopoulou of the Greek Ministry of Culture, had surveyed caves and rock shelters near the mouths of freshwater streams, the sorts of sites where Palaeolithic man is likely to have lived. Professor Curtis Runnels of the Plakias survey team told the newspaper: '[They] reached the island using craft capable of open-sea navigation and multiple journeys – a finding that pushes the history of seafaring in the Mediterranean back by more than 100,000 years.'

All in all the team recovered more than 2,000 stone artefacts from 28 sites. The quartz rocks from which the tools were fashioned were sufficiently abundant for tools to be discarded when blunted. At five of the sites the geological context had helped the team arrive at an approximate age for the stone tools. Professor Runnels estimated that they were at least 130,000 years old and could be much older.

In the report Dr Runnels suggested that they could be at least twice as old as the geologic layers (i.e. 260,000 years old). Dr Strasser went further – they could be at least 700,000 years old. Dr Strasser, who has conducted excavations in Crete for the past twenty years, bases his estimate on similar double-bladed hand axes fashioned in Africa 800,000 years ago. He believes that the large sets of hand axes found on the island suggest a substantial population, who must have made multiple sea crossings, all landing at the same place. In order to achieve that, Palaeolithic man must have had seaworthy vessels more than 100,000 years ago. He must also have had a system of navigation which allowed later arrivals to find Crete and the landing stages reached by earlier voyagers. The newer settlers must have heard about the previous voyages in detail – this at a time before the invention of writing as we know it.

Before these extraordinary finds were made, it was thought seafarers had not reached Crete until around 6000 BC. But it is now generally accepted that by 4000 BC at the latest Cretan people had

91

sufficient food, shelter and clothing to have surplus time on their hands – and at this point life ceased to be purely a battle for survival.

The Plakias survey team believed that this was the earliest evidence of Lower Palaeolithic seafaring so far found in the Mediterranean. Crete's very first inhabitants had arrived at the island by ship. Who knows what archaeologists will find next? We don't know in what sort of craft, but it seems that people were navigating Homer's 'wine-dark sea' tens of millennia earlier than anybody has ever dared to think.

10

LIFE IN THE LIBRARY

My task now, if I was going to get any further with my quest, was to find out more about the Minoans' need to sail. What exactly were they trading and how far did they have to travel to get it? The Uluburun wreck had told a fascinating story and I needed to unravel it. Dauntingly, I had to bridge the more than 3,000-year gap between the wreck and our own age.

Luckily it was not that difficult to stroll from my home in north London to the British Library. I spent three long weeks researching the Uluburun wreck and the objects found in its hold, going back and forth to the library's enormous new red-brick building on the Euston Road, breathing in the calm and scholarly atmosphere with real enjoyment. It is somehow rather comforting to sit inside a building with so much knowledge propping it up from all sides. It feels like a cocoon against the world, a time shuttle carrying every-thing in its red-brick hold, from beautiful historic maps to rare manuscripts. I like to think of all those books and papers mouldering away in basements underneath but the reality, as a diligent and helpful librarian explained to me, is that they are kept in vast climatically controlled racks underground, often in off-site storage facilities as far away from London as Boston Spa.

Now I was able to check up on the Uluburun's cargo in detail. Only a few months ago, the ancient world had been simply darkness, as far as I was concerned. Gradually, flashes of bright colour were piercing through that shadow and murk. I was beginning to get a feel for the real lives of a fast-developing people who had been inhabiting an extraordinary, active, bustling world that a few months before I had scarcely known existed. I also wanted to discover more about the Egyptians' political and strategic alliances. It seemed to me that the Minoans must have developed a long-standing relationship with Egypt, if they'd traded so many goods from Africa and the Middle East.

Fairly quickly, I learned that Minoan Crete had been renowned for its sophisticated perfumes, which it exported across the Mediterranean. Terebinth, with its hyacinth-shaped candles of bright red flowers, still grows on Crete – and is often used to lace the local brandy. To people in ancient cultures scent was of great value. They burned many kinds of resins and woods in their religious rituals, as well as adorning themselves. I already knew from past travels to Egypt and the west bank of the Nile that Queen Hatshepsut had led whole expeditions in search of another intense scent, incense: the results were recorded on the walls of the mortuary temple created in her honour at Deir el-Bahri, near the entrance to the Valley of the Kings. In the Uluburun wreck's hold the spirit of 'waste not, want not' was evident – the stalks and leaves of the terebinth bush itself had also been used on the ship as a springy packing material, protecting delicate items on the journey.

One look at a botanical tome, and unwittingly the humble terebinth plant had helped me to find what I had not quite known I was looking for. The resin only comes after the early winter frosts. This small fact, along with the fact that the ship was also carrying ripe pomegranates in the hold, stopped me in my tracks. Surely, this meant that the ship was on a winter voyage? This would imply that by the time the Uluburun sank, and maybe long before that, voyages were no longer confined to high summer. Bronze Age sailors were sailing in all seasons, and into the wind.

These early research successes aside, I soon realised that the task I had set myself was not as straightforward as it might have seemed. There was a geographical anomaly in the Uluburun's hold, and it was staring me in the face.

There was a rare ceremonial sceptre-mace on board. The expert analysis was that it came from either Bulgaria or Romania – from halfway up the Black Sea. Also in the hold was a large quantity of amber, the fossil produced by pine resin, which actually came from the Baltic. I thought at once of Severin's voyage – the parallel with Jason's voyage and the tale of the Golden Fleece – and then brushed it from my mind. This was real life we were talking about, not myth. Tim Severin had got as far as the Black Sea. The Baltic was an altogether more dangerous proposition. Had the Minoans truly obtained Baltic amber all the way from the north? Today polished amber is highly prized, treated like a jewel and bartered like one. But then?

A quick look at the work of the great classicist Robert Graves[6] and some science journals yielded a further series of supporting clues: apparently, wine is quite possibly a Cretan invention and not just in myth. The linguistic evidence comes with the fact that the very word is from Crete. The vines used to make it are thought to have first come from the area around the Black Sea, where they originated in the wild.

My research also confirmed that amber had been both valued and used since the Stone Age. The Greek name for amber is *electron*; the origin of our own word electricity. Rub amber with a cloth and it will become electrified and even attract paper.[7] It must have induced a real sense of mystery and wonder. It was certainly used in sympathetic magic. Worn as a charm, the belief was that amber protected against sore throats, toothache and stomach upsets. It was also seen as a remedy for snake bite and newborn babies were given amber necklaces to protect against infection. In much later times, even the prophet Muhammad valued it: he said that a true believer's prayer beads should be made of amber.

So amber was a true treasure; a great prize. To obtain it our ancient mariners would have needed to sail through the Kattegat and Skagerrack. Even those great sailors of latter times, the Vikings, had occasionally come to grief doing that.

I cast my mind back to Jason and the Argonauts and thought how dangerous travel in rocky seas would have been. Yet the Baltic amber that had evidently been found in many ancient gravesites in mainland Greece, as well as on the Uluburun wreck, must have got there somehow. Neither did it feel like a coincidence that Crete had such a reputation for wine, if its sailors had been the first to discover vines, and grapes, because of their adventurous travels.

I could see that to piece all of the strands of evidence together I needed to learn much, much more about the fascinating process of world trade from all that time ago. I decided to concentrate on four significant categories of the traded goods and see if I could discover more about the process. Taking a notebook, I began what for me was several days' work, but it is summarised here as just five sections.

AMBER

Infrared spectroscopy proves that most of the amber found in ancient sites in the Mediterranean came from the Baltic, as did that in the Uluburun wreck. The earliest Baltic amber found in the Mediterranean was found at Mycenae (shaft grave O) and is dated to c.1725–1675 BC. A vast amount of the jewellery found at Mycenae was originally Minoan.

A total of 1,560 pieces using amber were found at Mycenae, 1,290 from shaft grave IV alone. It is intriguing that these earliest consignments show a remarkable similarity (in the design of their spacer plates) to amber necklaces of the same period found in Britain. Which also originated in the Baltic. So it appears possible that Crete's expanding empire was already trading in amber by 1725 BC.

I soon discovered that there are others apart from myself who believe that the Minoans were behind the ancient amber trade. In 1995 Hans Peter Duerr, who was then a director of the German scientific research organisation the Max Planck Institute, decided to take his family on holiday to the North Sea islands, near Hamburg. Duerr was interested in the lost city of Rungholt, which in the Middle Ages sank beneath the waves in a tempest.

Medieval Rungholt, then a part of the nation of Nordfriesland, had a population of some 1,500 people and a reputation as an astoundingly wealthy port. It became victim to the first 'Grote Mandraenke', a Low Saxon term for 'A Great Drowning of Men' – a vicious North Sea storm tide that swept over England, Denmark, Saxony and the Netherlands in 1362, smashing island groups apart and killing an estimated 25,000 people.

Duerr's most exciting findings, he told the German magazine GEO, lay underneath the late medieval Rungholt – beneath, and therefore older than, a Bronze Age layer of peat dated to 1200 BC. They were just ordinary, everyday items; items that sailors, who were there for far more valuable goods, had left behind at a port.

> We came across remains of Levantine and especially Minoan ceramics, the daily kind used to transport goods. They were dated 13th and 14th century BC. Amongst these were shards of two tripod cooking pots from Crete. That's why we believe ships were sailing in 1400 BC from Crete to the coast of Northern Frisia.[8]

There is a possibility that these pots could have been antiquities which were being carried much later by a more modern ship. Duerr doesn't believe that, because the items themselves were so ordinary. The cooking vessels were scarcely 'antiques'.

> The pots we found were not trade goods being carried by merchants. The ceramics were of practical daily use – belonging, with great certainty, to equipment on a ship.

The finds did include lance tips and incense, but mainly what they found were containers for drinking and eating. Most crucially, Duerr discovered a seal that he claims had a Linear A inscription on it.

What could have tempted the Minoans from Crete to the North Sea in the 14th century BC?

Their interest was first prompted, Duerr believes, by: 'Tin from Cornwall. The finds point to a shipwreck.' He went on:

> I can now add it was not far from Britain to Frisia, where amber came from, beloved by the people of Mycenae. It was possible for the Minoans to navigate the North Sea 3,300 years ago . . . .

Tin from Cornwall: navigation of the highly dangerous North Sea . . . For now, I was going to put those astonishing assertions aside and continue investigating the Uluburun's hold.

ELEPHANT TUSKS

A length of elephant tusk, cleanly sawn at both ends. Heavily stained by the copper ingots in the wreck, it was found at the stern of the ship. Ivories were regularly traded as luxury goods. I vividly remembered seeing the charred ends of an elephant tusk in Heraklion's museum; it had been found in a 15th-century BC tomb at Zakros in Crete. Elephants are, of course, found in both Africa and India, but the tusks discovered in the Uluburun wreck have not yet been classified. However, we know Indian elephant tusks did reach the wider Mediterranean in the Bronze Age. Five tusks from Indian elephants have been found in the ruins of a Middle Bronze Age palace at Alalakh in southern Turkey and more at Megiddo, in northern Israel's Jezreel valley.

HIPPO TEETH

I have a soft spot for the hippopotamus and its teeth. In 1959 I had the misfortune to be serving on HMS *Newfoundland*, a cruiser returning from Singapore to the UK to decommission. *Newfoundland* was an old Second World War ship, useless in the age of strike aircraft and nuclear submarines. She had a complement of fifty-eight officers, when eight would have been enough. For most of the time we had nothing to do, so we drank heavily to pass the time.

We berthed at Lourenço Marques in southern Mozambique and spent the first night at a seedy Portuguese night club. One of my favourite singers, Maria de Lourdes Machado, was singing *fados*, the sad Portuguese songs of lost love and betrayal. I think we must have all over-indulged . . . It was a beautiful dawn and we decided to go on a hippopotamus shoot in the delta of the Limpopo River, to the north. We took one of the ship's motor boats, a crate of rum and some limes and set off.

We found plenty of hippos snorting in the shallows and we started shooting at their flat backs. The hippos didn't like this at all. In fact, they rammed the boat. I remember seeing the boat in the air, cartwheeling upside down, with the propellers still turning. An angry hippo with red, bloodshot eyes and huge teeth was staring at me from a few feet away, eyeing me up for breakfast. Too drunk to care, I vomited green bile, which spread in an oily slick across the muddy Limpopo water towards him. In disgust, he dived under me and disappeared.

So I had had a good enough close-up of hippopotamus teeth. More to the point, their teeth have been worked into ivory jewellery for a very long time. A fragment of a hippopotamus lower canine was found at Knossos in the ruin of an early Minoan (3rd millennium BC) palace. So there was trade between Africa and Crete long ago.

SHELLS

It's extraordinary to think that objects as delicate as ostrich eggs managed to survive the wreck of the Uluburun, but some of them did. At least five tortoise carapaces were also on the ship; the bowl-like shells were used for making lyres.

But I was interested in two types of shell in particular. The first sort came in their thousands: those of *murex opercula* sea snails. Crete was the world centre of the trade in the prized purple dye that was extracted from these incredibly smelly molluscs. It took thousands of murex to make enough dye for the hem of a cloak and the Minoans had farmed them in great numbers for this lucrative trade. The presence of so many of them on the wreck supports the idea that the ship was Minoan.

I was also intrigued by twenty-eight rings from an unidentified, large shell. The rings were found cut into shape and ground down. Their size implies that they were not from the Mediterranean, but from the Indo-Pacific region. They had already attracted the attention of experts:

> ... The Uluburun rings provide evidence for trade between the Persian Gulf and the Levantine coast during the 14th century BC. Shells were either imported into Mesopotamia as finished rings, as may have been the case at Usiyeh, or made into rings there and probably also embellished with inlays affixed with Mesopotamian bitumen, before being exported to the Levant.[9]

So here we have a hint of the trade between the Minoans in the Mediterranean and the Indian Ocean.

THE COPPER INGOTS

When he discovered the Uluburun wreck Mehmet Cakir was looking for sponges. What he found, instead, he said, were biscuits – 'biscuits with ears'. What looked to him like biscuits turned out to be copper and tin ingots, lying near the keel. Many of the tin ingots had corroded into sludge, but the copper ones remain in remarkable condition even after their 3,500 long years under water.

The total weight of the ingots was some 11 tons – 10 of copper and one of tin. There were 354 copper ingots in the shape of an oxhide, 121 smaller bun-shaped copper ingots and fragments forming another nine. No moulds have been found to show how the molten metal was formed into these shapes, but it seems that there were two pours of molten metal into one mould, in rapid succession – evidenced by cracking during cooling as the metal contracted. Most of the ingots were incised with marks when cold – probably at the trading place where they were collected and sold.

The origins of this copper are heatedly debated: this is also true of the tin. If we knew its sources, it could explain how the Mediterranean exploited such enormous quantities of bronze, when there appeared to be insufficient numbers of mines to satisfy the demand.

A thorough analysis of the copper ingots has been carried out by Professors Andreas Hauptmann, Robert Maddin and Michael Prange. It is described in their paper 'On the structure and composition of copper and tin ingots excavated from the shipwreck of Uluburun'.[10] They write:

> The ship carried ten tons of copper and one ton of tin. The cargo thus represented the 'world market' bulk metal in the Mediterranean . . . Cores drilled from a number of ingots show an extraordinarily high porosity of the copper. Inclusions of slag, cuprite and copper sulphides suggest the ingots were produced from raw copper smelted in a furnace and in a second step re-melted in a crucible. Internal cooling rims point to a multiple pouring. We doubt that the entity of an ingot was made of one batch of metal tapped from a late Bronze Age smelting furnace. The quality of the copper is poor [viz. the smelting process] and needed further purification before casting, even if the chemical composition [i.e. the raw copper] shows that it is rather pure. The copper was not refined. The tin ingots in most case are heavily corroded. The metal is low in trace elements, except for lead.

The authors were studying the smelting process. Yet in examining that procedure, they also had to analyse the raw material. And the results intrigued me. They continued:

> From the chemical point of view, the purity of this copper is extra-ordinary in comparison with other sorts of copper distributed in the late Bronze Age Old World. For instance copper from the Wadi Arabah is much higher in lead (up to several per cent) . . . copper from the Caucasus area is extraordinarily high in arsenic (up to several per cent) . . . copper from Oman usually contains arsenic and nickel in the percentage area.

The authors don't believe that this extraordinary purity could have been the result of smelting

> . . . the concentrations, for instance, of lead, arsenic, antimony, nickel or silver do not change very much during smelting . . . We therefore

conclude that the ingots reflect the composition of 'pure' copper ores that were smelted to produce the metal.

Detailed results, including a table for the composition of each ingot, showing the staggering purity of the copper, are contained on our website.

I was more than surprised. There is only one type of copper with that level of purity, the copper that comes from Lake Superior on the Canadian–American border. I only knew about it because many North American readers of my earlier books about Chinese discovery, *1421* and *1434*, had written to me on the subject. The Keweenaw Peninsula in Michigan still boasts some of the purest copper ever found: a metal so pure you scarcely had to refine it to make your burnished copper cooking pots. Millions of pounds of copper from North America – mined in the 2nd millennium BC – appeared to have been exported somewhere, no one knew where. Readers had been wondering if it had been taken back home to China, in Chinese ships.

How could ten of the bun-shaped copper ingots found in the Uluburun wreck be made up of Lake Superior copper? But then, how could a tiny American tobacco beetle have turned up in a ruined merchant's house on Thera, in the middle of the Mediterranean?

In an Annex on our website there is, firstly, Professor Hauptmann and colleagues' report on the chemical analysis of the Uluburun wreck copper ingots. Secondly, there are extracts of thirteen reports on Lake Superior copper. As may be seen, all thirteen copper samples from Lake Superior and ten ingots from the wreck have purity of at least 99 per cent – purity unique to both Uluburun ingots and Lake Superior copper.

The great pyramid of Khufu, Egypt.

An image, painted circa 1479–1425 BC in the tomb of Menkheperraseneb, clearly shows a 'Keftiou' (Cretan) bearing a gift of a bullhead rhyton.

A view of the magnificent Palace of Knossos, Crete.

The archaeologist Arthur Evans, who first unearthed the palace of Knossos and named the 'Minoans' after King Minos.

The bullhead rhyton unearthed at Knossos.

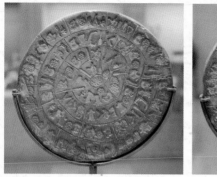

The enigmatic Phaestos disc. Experts have struggled for over a century to decipher both it and the mysterious Minoan language, Linear A.

A view of the great Palace of Phaestos, Crete.

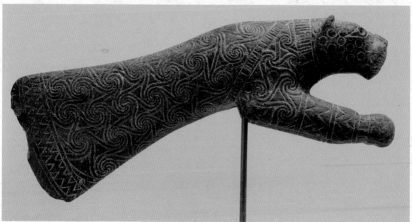

A beautiful carved axe in the form of a panther – Heraklion Museum, Crete.

A beautiful Minoan fresco at the Palace of Knossos.

The throne room at Knossos.

Pithoi storage jars, as seen at Knossos, have been found at various sites around the Mediterranean and beyond.

Jewellery in the 'Aigina Treasure' – the Master of Animals at the British Museum.

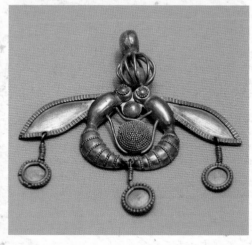

A golden bee pendant and a Minoan bee brooch at the Heraklion Museum, Crete.

The flotilla fresco at Akrotiri,
unearthed by Dr. Spyridon Marinatos in 1967.

Items from the Uluburun shipwreck treasure trove include copper ingots, hippo teeth and elephant tusks.

Bronze tools and implements from the Yemeni Al-Midamman bronze hoard.

That evening, unable to put the subject or myself to bed, I read up on the investigating chief archaeologist's account of the find. Dr Cemal Pulak is director of research at, and vice-president of, the Institute of Nautical Archaeology in Turkey.

Because of the enormous value of the goods in the Uluburun's hold, Dr Pulak believes that when it came to grief the wreck had been carrying a royal or élite shipment. Even some of the 'everyday' objects on board bear this theory out. For instance, both the Minoans and their successors the Mycenaeans preferred to be clean-shaven. Dr Pulak believes that at least two Mycenaeans were aboard, escorting the goods. His theory was partly based on the evidence of five typically Mycenaean bronze razors found on board.

Dr Pulak's thesis is that the ship's home port may have been in Canaan, in what's now the north Carmel Coast of Israel, mainly because of the characteristic Canaanite design of twenty-four stone anchors found with the wreck. However the typically Minoan construction of the ship leads me to think that its origin was on either Thera or Crete, during the Mycenaean era.

The ship's carrying capacity was at least 20 tons, the archaeological team calculated, a figure they had worked out by tallying the recovered objects, including the 10 tons of copper and tin. I had estimated the admiral's larger vessel's carrying capacity at 50 metric tons.

|||||||||||||||||||||||||

It was from Dr Pulak's work that I learned, with great curiosity, of the existence of the so-called 'Amarna Letters'. The name Amarna had cropped up before, when we were on Crete. I now knew it as the alternative capital city founded by the Egyptian pharaoh Akhenaten of the late 18th Dynasty – husband to Nefertiti – and that the new capital had been abandoned shortly afterwards. The letters prove that during Akhenaten's reign there was a sophisticated and developed system of trade going on between Egypt and a number of nations. They also show just how crucial it was to Egypt's power base to keep up a continuous supply of bronze.

Akhenaten (c.1353–1336 BC) is now remembered mostly as father to Tutankhamun, the famous boy king. Statues show Akhenaten to have been fleshily handsome: his long face bears large, well-defined features and has a high forehead. In fact, you might even say he looks headstrong. Akhenaten was a radical, so much so that he broke with the worship of traditional Egyptian gods. He built his new capital Amarna in an extraordinary new style. Thousands of years later, in 1887, a Bedouin woman was working in what appeared to be just a field when she discovered a cache of cuneiform (the ancient Mesopotamian lettering system) stone tablets.

Dating of the 382 tablets is difficult, except to say that most of them would have been written well before Amarna was abandoned, shortly after Akhenaten's reign.

As the translator and Assyriologist William Moran confirms: 'The chronology of the Amarna letters, both relative and absolute, presents many problems, some of bewildering complexity, that still elude definitive solution.'[11]

What these tablets and fragments do tell us is about the established and elaborate trading etiquette that existed between kings. For example, the king of Alashiya wrote that he was late on delivery because much of his workforce had been 'slain' by the god of pestilence:

> As to the fact that I am sending you only 500 [shekels of] copper, I send it as a greeting gift to my brother. My brother, do not be concerned that the amount is so small. In my country all the men have been slain by the hand of Nergal, and there is nobody left to produce copper ... send me silver in quantity, and I will send you whatever you request ... [12]

Experts think this was a negotiating tactic: if the king can send copper later, then it's unlikely that the men he would have relied on to produce it had genuinely died. In effect what the king is doing is buying time and testing out his potential client: he is sending a sample and asking Akhenaten for silver in return.

This world was beginning to form more and more complex

dimensions; a multi-faceted but elusive jewel, the colours drifting slowly into place. To me, it looked increasingly as if the Minoans had sailed to all parts of the compass. It also looked like I was going to have to follow them.

11

A PLACE OF MANY NAMES
AND MANY NATIONS

Today's destination: Tell el Dab'a, a Middle Kingdom palace on a hill in the Nile Delta. Our goal is an ancient port beside the modern city, a place that was named Avaris during the Egyptian 13th Dynasty, when it was a crucial trading port dominated by the commercial traders known as the Hyksos.

A group of friends had agreed to get involved with the first stage of my quest: an expedition to Tell el Dab'a and then on to the Nile Delta, following the old Red Sea–Nile canal north of Cairo to Zagazig, to test what I thought must have been Minoan trade routes. *En route* we visited Bubastis, the old capital city of the 'Cat Pharaoh', Bastet, simply for fun, but my true aim was to track Minoan influence and involvement though ancient Egypt.

I'd decided that I would have to explore the routes the Minoans had taken, using the sources of the Uluburun wreck's treasures as my guide. Doing so would be a £100,000 gamble of a trip starting in Beirut, then on via Damascus and Aleppo to Babylon, the beating heart of the Bronze Age world. All were important trade hubs during the Bronze Age. But first, I was going to take my time travelling through one of my first loves, the sacred land of Egypt, hunting down the truth behind this statement, written by Bernard Knapp:

... Egyptian Keftiu documents clearly indicate the leadership of Crete ... The recently uncovered 'Minoan style' frescos at the site of Tell el-Dab'a in Egypt's eastern Delta and the Minoan style painted plaster floors at Tel Kabri in Israel open up the likelihood of diverse social or political contacts.[13]

It can be much harder to locate and investigate the remains of Bronze Age Egypt here in the fertile flat lands of the Delta north of Cairo than it would be in the dry south of the Nile valley. For millennia the Delta has bestowed plenty upon Egypt. These are some of the richest agricultural lands in the world: Egyptian farmers simply plant their crops and wait for the annual Nile floods to fertilise and water their land. Naturally, this fertility has meant that the land has been ploughed up mercilessly. But the wetter atmosphere has also made the old cities of the Delta crumble, while the great works of art of the old civilisation have long since disappeared back into the soil. Ashes to ashes.

To begin with, we simply munched croissants and sipped strong Egyptian coffee, watching farmers set off to work on their donkeys. Skeins of geese migrate in the sky high above us. Each house has its dovecote. Now and then an old railway train chugs across the flat land, belching black smoke over the people hanging from its roof.

It was market day and the village was full of women in black burkas carrying their shopping on their heads. Their baskets look like a cover for one of Claudia Roden's beautiful Middle Eastern cookbooks: loaded down with beans, cucumbers, lentils, chickpeas, dates, olives and cabbages.

But personally I've been drawn here not by the delicious food, but by a geographical survey by the Austrian Archaeological Institute. The survey revealed the existence of a buried harbour basin about 450 metres square (540 square yards), with a canal connected to the Pelusiac branch of the Nile.

This region is where the biblical Jacob is said to have brought his family. All that's left to see today are a few flattened, dusty ruins in the farming villages round about. The rest, like much archaeology, stays firmly hidden underneath the modern town. But in c.1783–1550 BC this whole area must have been a hive of commercial activity. Radar imaging by the team of Austrian archaeologists shows that in the Bronze Age not only were there two islands and a tributary of the Nile running through Avaris, but a second harbour once sat alongside Palace F/II of the Middle Hyksos period. Historians have always thought that Egypt was no sailing nation. If this is true, then why the harbours? A third harbour or dry dock lies north of Avaris, at the Nile branch itself.

Bronze Age Egyptians are probably best known for their remarkable building achievements and their elaborate cult of the dead, especially the art of mummification. Until very recently, it was assumed that the Egyptians knew little of boatbuilding, despite the obvious importance of the Nile. Yet after the Hyksos period, Avaris became a famous naval base. It was built primarily by Thutmose III and Amenhotep II and was at times called Peru-nefer.

Timing is difficult to establish, for certain, and it involves a degree of analysis of the complex knots of peoples, races and historical figures who have passed through this ancient land. The Hyksos probably arrived in the late 12th Dynasty (Middle Kingdom) period. They may have come originally as shipbuilders, sailors, soldiers and craftsmen. It's not difficult to imagine Avaris as the Dubai of its day, with a vast building workforce drafted in from overseas. The pharaohs settled them here deliberately in the late 12th Dynasty, to create a harbour town and perhaps even build the ships. But at a later time of political weakness the workmen established their own small but independent kingdom and had to be swatted back. Hyksos artefacts have been found in the Knossos labyrinth.[14]

Thutmose III, also known as Thuthmosis, was the warlike stepson of the bearded lady ruler Hatshepsut. He created the biggest empire Egypt had ever seen, an empire that ran from Nubia to Syria. At Luxor, on the heavily decorated walls of the tomb of Thutmose's valued vizier, Rekhmire, is a famous fresco. It shows a procession of men whose looks and dress are definitely Minoan (see first colour plate section). They come, as they say, bearing gifts. The painting is surmounted by a line of hieroglyphs. The translation reads: 'The coming in peace by Keftiou chiefs and the chiefs of the islands of the sea, humbly, bowing their heads down because of His Majesty's might, the king Thuthmosis III.'[15]

Whatever this city's name was – through time it has been Avaris, Piramesse or Peru-nefer – it was certainly a major port, bustling over the summer trading season and humming with the activity of many ships. And as I was about to find out, the Minoans, or Keftiu, were here in force.

Avaris/Peru-nefer became a crucial military stronghold. The city was the starting point for the overland route to Canaan, the famous 'Horus Road'. Known in the Bible as the 'Way of the Philistines' (Exodus 13:17), the road was used for military expeditions as well as for commercial traffic. The site appears to have been abandoned for a time, after the Hyksos were driven out. However, by the end of the 18th Dynasty, when the Egyptians were back in control, Avaris boasted three large palaces ringed by an enclosure wall. The whole complex was about 5.5 hectares (13 acres) in size. At least two of the palaces excavated here, Palace F and Palace G, held some truly extraordinary finds.

When the Austrian archaeologist Professor Manfred Bietak first worked here in the 1960s, he was amazed to come across thousands of fragments of exotic wall paintings. They did not look remotely Egyptian. As he pieced them together, a somehow familiar work of art gradually unfolded before his eyes. It had a beautiful blue background. As he worked he realised that there were some human figures: one was jumping. He was leaping over a bull. The archaeologist was astounded. He had definitely seen that image before. But where?

In fact Bietak's boy – seen jumping over bulls on a blue background – was exactly the same image as that on a fresco unearthed by Arthur Evans, almost a century before. At Knossos. The electrifying discovery that Minoan artists had worked at foreign courts has had the art and museum worlds transfixed ever since. And now me.

A number of different scenes showing bull hunts and acrobatics were discovered here, painted on the hard lime plaster favoured by the Minoans of Crete, rather than Thera. Some of them are set against a maze pattern. Hunting scenes, life-size male figures with staffs and heraldic griffins have been pieced together from the tiny fragments of plaster scattered across the site. The griffin is a typically Minoan motif and the Avaris griffins were as large as the ones painted at Knossos. There is also an intriguing painted figure of a woman, shown in a flowing white skirt similar to those of the priestesses of Knossos. Although it is impossible to establish with absolute certainty whether the patrons and the painters were from Crete, this is what the experts now believe. It looks strongly as if this was a Minoan trading outpost, a wing of the widespread empire I now suspected existed.

If we think about today's modern, multi-billion international trade in art, the thought prompts images of high fashion, of status, of rich people striving to create an ultra-sophisticated world that will impress those around them. The Minoans' great artistic skills would have been in great demand, because they were of the highest order and were very rare in the Bronze Age world. Yet the colourful frescoes of Avaris are unlikely to have just been items of fashionable interior decor. On Crete, artists' use of the bull-leaping scenes and the half-rosette symbols were tightly controlled. Many of the Tell el Dab'a scenes show motifs which the experts believe are specifically royal: they may also have had spiritual significance. The images were restricted to formal buildings, particularly at Knossos, indicating the city's power. This outpost, therefore, must have had the same command and control-type role.

It's more than possible that a political encounter on the highest level took place between the courts of Knossos and Egypt.

In Harold Evans' rather old-fashioned, Edwardian take on the world, the throne room at Knossos was created for the male ruler, Minos. Now, after nearly a century of new finds and new palace digs, archaeologists think that the throne was made not for a king but for a queen. The Minoans' gods were female. As in Egypt, the ruler was the gods' representative on earth. At Knossos, that queen sat on her throne between two seated griffins – an allusion to her role as the Great Goddess, Mistress of Animals. Those griffins are exactly like the pair uncovered at Avaris, in Palace F.

Could it be that the Minoan presence in Egypt was formalised in the time-honoured way, by marriage? Having this base would mean that the Minoans could not only trade *en route*, but provision themselves for a long voyage. They could load their ships with dates, fresh vegetables and salt fish for expeditions to the East.

Digs at the surrounding settlement, dated by pottery and scarabs, place the palaces in the late reign of Thutmose III and that of his son, Amenhotep II. This fits in with the scenes on Thutmose's vizier's tomb. It all made perfect sense. A formalised pact with the Minoans would have provided a pathway for Egyptian traders to the southern Greek mainland and Greater Anatolia. It would also have given the Egyptians what we would now call in our modern-day terminology 'knowledge transfer' – access to the Minoans' tremendous seafaring and shipbuilding abilities.

Perhaps there was more besides. I'd arrived armed with the knowledge that the art world, at least, was convinced that even

before the birth of Christ the art market was fully international. Gradually, I was forming a picture of the Minoans as trendsetters as well as globetrotters: the Damien Hirsts of their time, skilled practitioners who possessed ingenuity beyond price. After all, at Thera they had painted graceful and lifelike scenes of monkeys, antelope and lions: paintings so accurate that they had to have been copied from life. Minoan artists were not just skilled. They were celebrated.

So in a sense it was no surprise for me later to find that their work appears on ancient palace walls not just here in Egypt but in a huge, south-sweeping arc through the eastern Mediterranean – at Mari in modern-day Syria, at Ebla about 34 miles (55 kilometres) away, at the ancient royal town of Qatna, which is currently being excavated by a German/Syrian alliance, at Alalakh in southern Turkey and at Tel Kabri in Israel. The ancient city-state of Alalakh is near Lake Antioch, in the Orontes River valley. Sir Leonard Woolley, who led the excavations in the 1930s and 1940s, uncovered royal palaces, temples, houses and town ramparts here. The discovery of what we would now interpret as Minoan frescoes came slightly later. At that time, archaeologists hadn't built up an understanding of the Minoans' cosmopolitan adventures, or of the no doubt complex relationship of Crete with Anatolia. The discovery of the frescoes came more or less out of the blue and threw the art world into consternation. The revelation caused archaeologists to speculate that Crete's magnificent murals were in fact Asian. Woolley argued that:

> There can be no doubt but that Crete owes the best of ... its frescoes to the Asiatic mainland. We are bound to believe that trained experts, members of the ... Painters' Guilds were invited to travel overseas from Asia, to ... decorate the palaces of the Cretan rulers.[16]

Yet the evidence that these were Minoan artists of great skill, exporting their talents in an arc reaching all the way to Babylon, has mounted inexorably with each new dig in the region. The excavators of Tel Kabri discovered fragments of a fresco that looked almost exactly like a reproduction of the beautiful miniature fresco in the West House on Thera.[17]

Written archives found at the ancient Sumerian city of Mari show that King Zimri-Lim, who lived between 1775 and 1761 BC, prized his Minoan pieces so highly that he gave them as prestige gifts to other rulers. The poetry of Ugarit, an important city of the time that is today known as Ras Shamra, suggests that the Minoans didn't just decorate foreign palace walls. They were architects of international renown, who built their clients' buildings too.

IIIIIIIIIIIIIIIIIIIII

Before I'd left for Egypt, the Metropolitan Museum of New York had mounted an exhibition named 'Beyond Babylon'. The fascinating catalogue was written by a series of specialists, their expertise drawn together specially for the exhibition.[18]

The experts believe that the discoveries at Tell el Dab'a testify to an Egyptian special relationship with Crete. As Joan Aruz wrote in the catalogue:

> The stunning discoveries of unquestionably Aegean-looking frescos around the Mediterranean littoral have dramatically enhanced our picture of cultural exchange during the 2nd millennium BC ... the presence of Minoan artists at foreign courts has transformed our view of cultural interaction in the eastern Mediterranean world.

'Beyond Babylon' also cites 'Papyrus BM 10056', a document in the British Museum that mentions Cretan ships being docked in an Egyptian harbour. The document refers to a place named 'Perunefer'. The palace complex at Avaris/Peru-nefer had evidently had quite a history: Tell el Dab'a may well at one point have been home to the biblical figure Moses, as well as being the summer residence of the pharaohs. In short, prior to the Exodus of around 1446 BC, it was one of the most exciting places in the world; a totally cosmopolitan melting pot and a meeting place for world leaders. David O'Connor, Professor of Egyptian art and archaeology at New York University, wrote the following in the catalogue:

> The usual interpretation is that these ships, which are mentioned only during the reign of Thutmose III, were of Cretan type, or were sailing to Crete. However, it would be more logical to assume that Minoan ships were actually moored, and were repaired at Perunefer. If one can identify Tell el-Dab'a, with its palatial Minoan wall paintings as Perunefer, then it is conceivable that Egypt

fostered its special connections with the Minoan Thalassocracy in order to build up its navy for military enterprises in the Near East.

If Palace F was indeed a Minoan political base placed right within the beating heart of Pharaonic Egypt, the most powerful nation on earth, my case for a Minoan super-trading empire – beginning with this truly special strategic relationship – was getting stronger all the time. Avaris, with its repair facilities, food and water, in effect provided an ideal forward base for onward voyages. So – exactly how far across the world did the Minoans get?

12

A SHIP IN THE DESERT

My friends' flight home was in a few days: I was due to go on to Mari. In the meantime, while I was still in Egypt I got a call from Marcella, already dutifully back at her work desk in London.

By a stroke of luck, she had come across a report in *USA Today* that shed more light on my quest. It was a piece by Dan Vergano, the magazine's science reporter, about a new discovery in the desert. She read the report out to me:

> Archaeologists generally downplay the *Indiana Jones* side of their discipline, full of derring-do and unexpected discoveries. But every once in a while, an amazing find surprises even the most experienced researchers ... That's just what happened when Boston University's Kathryn Bard reached into a hole in the sand at the edge of the Egyptian desert ... Her research team of Italians and Americans now knows those caves hold the most ancient ship stores ever discovered; perfectly preserved timbers, ropes and other fittings perhaps 4,000 years old.[19]

Braving high temperatures and the poisonous vipers that are rife in the desert, in December 2005 Bard had found a hidden chamber in an area named Wadi Gawasis, along the Red Sea coast. Exploring the back of a cave, Bard's fingers had met with thin air; making the team realise that there was a hidden chamber of some kind, waiting to be found. Later, her Italian colleague Chiara Zazzaro cleared some fallen rock – and exposed the back of a second cave. The cave had been expressly cut, by hand, from the rock. Here, on this dried-up ancient watercourse, the team had found a hidden, secret shipyard.

Indiana Jones associations aside, this was genuinely an extraordinary moment. Finding an industrial site, one that tells us where everyday Egyptians worked, rather than a carefully preserved ceremonial one, is unusual enough. But finding a site like this, with many working materials still intact and untouched after perhaps 4,000 years, is absolutely unprecedented in Egypt. The amazed team opened up cave after cave and found ancient coils of rope, ship parts, jugs, trenchers and everyday linens; all deeply practical items which drew a lively real-life picture of Egypt's ancient seafaring past. To date, the team have uncovered a complex of eight caves, a network of rooms filled with relics over 4,000 years old that proved that the Egyptians had mastered advanced ship technology. In the complex were dozens of nautical artefacts: limestone anchors, eighty coils of knotted rope, ship timbers and two curved cedar planks that seemed to be the steering oars from a 21-metre-long (70 feet) ship.

Bard and her colleagues now believe, from studying satellite images at the Wadi, that there may be another ancient structure needing investigation, in the form of a slipway or dock below what was the Pharaonic harbour.

When they created this harbour, the Egyptians were almost certainly aiming to exploit the wealth of the famous Land of Punt. The actual location of this fabled place is a mystery, although Bard thinks it may have been in today's Sudan. The pharaohs were organised, methodical and they thought for the long term. Although such expeditions were probably a rarity, they must have been occasions of great prestige. Sometimes Punt is referred to in Egyptian records as *Ta netjer*, the 'land of the gods': they must have prized it highly. All we really know is that the mysterious Land of Punt, or Pwnt, was fabled for its prized luxury goods such as wild animals, perfumes, African blackwood, ebony, ivory, slaves and gold. It may have been in today's Somalia or around the hook of Africa in Ethiopia: but wherever it was, it was an entry point to the tremendous and exotic natural wealth of wider Africa.

We know that the most famous ancient Egyptian expedition that sailed to Punt was made personally by the remarkable Egyptian queen, Hatshepsut. She is not the only female ruler in Egypt's illustrious past, but she was certainly the only one regularly depicted wearing a beard. Hatshepsut built a Red Sea fleet to bring mortuary goods back to Karnak in exchange for Nubian gold. Details of her five-ship voyage to Punt are narrated on reliefs adorning her mortuary temple at Deir el-Bahri, mentioned in chapter 10. A voyage of this kind would have been a huge logistical undertaking, requiring scribes, quartermasters, pack animals, workmen and shipwrights, as well as sailors. This may have been the preparation point for one leg of her journey. Still, the new Wadi Gawasis find shows that enterprising Hatshepsut was by no means the first Egyptian ruler to target the riches of the land of the gods.

The archaeologists suspect that the port at Wadi Gawasis was used by the Egyptians for centuries; perhaps from as early as the time of the Old Kingdom (2686 –2125 BC) and lasting until around 1500 BC. They discovered limestone stelae in niches bordering one of the cave entrances. Many were indecipherable, but one clear-as-day inscription mentioned at least two early expeditions, one to Punt and one to Bia-Punt. They were commissioned by Amenemhat III (12th Dynasty), who reigned about 1860–1814 BC. The expedition was led by two brothers, one named Nebsu and the other Amenhotep. Another inscription found there romantically describes the sea the brothers set out to conquer as 'The Great Green'.

The craft appear to have been up to 21 metres (70 feet) long, powered by rowers and sail. The cedar timbers used to build the ships were, as you would expect, cut and aged in Lebanon, then shipped to Egypt. It looks likely that they were built into boats on the Nile, at a port site near modern Qift, then disassembled and trekked on donkeys across the desert for ten days and reassembled at Wadi Gawasis, which was then the site of a lagoon, long since silted up. Now the Institute of Maritime Research and Discovery is supporting the visit of several nautical specialists to the Mersa/Wadi Gawasis expedition.

It must have been an extraordinary moment when the archaeologists opened up this mysterious cave to find rope neatly coiled and knotted, stored exactly as some meticulous sailor left it, over 3,800 years ago. The team found forty large empty wooden boxes in the storage rooms – these were cargo boxes waiting to be packed up with exotic wares. Two of them were labelled with a painted inscription, like an advertising slogan. It read: 'The Wonders of Punt'.

13

NEW WORLDS FOR OLD

So: it was extraordinary to think of it, but I now had proof positive. Far from the Bronze Age being a dark and obscure time, with little going on except hunting, trapping and some farming, I had discovered an international jet-setting scene; a glittering art world and a sophisticated world market for metals and luxury goods. Avaris had been a revelation to me. The convention of history was wrong: the Egyptians had definitely ventured off their own shores, voyaging on ships that their honoured Cretan guests might have captained or crewed, like a modern lease. The Minoans and their sailing skills were so highly regarded by the Egyptians that they had been given special dispensation to set up an outpost there. With their skills in art, design and metallurgy, the glamorous 'Keftiu' sat at the very centre of this cosmopolitan world.

There was just one last enigma I had to solve to my own satisfaction, while I was still in Egypt. It was a small detail that took me back to the walls of Thera and to one tiny little American tobacco beetle that had been worrying me ever since.

In 1992 a well-respected pathologist, Dr Svetlana Balabanova, decided on an experiment. She took samples of hair, bone and soft tissue from nine Egyptian mummies and in a one-page article in the German publication *Naturwissenschaften* reported her astonishing findings of cocaine and hashish usage in all of the mummies. A further eight showed the use of nicotine.

Her findings were immediately attacked, on the grounds that two of the substances found – inside a mummy nearly 3,000 years old – were derived from indigenous American plants; cocaine from

Erythroxylon coca and nicotine from *Nicotiana tabacum*. Tobacco, which contains nicotine, is an American plant. The idea that there could have been any transatlantic contact between America and Egypt – not just before Columbus, but even before the birth of Christ – was obviously so ridiculous that the experts felt that actual scientific enquiry could be ignored.

Balabanova's team stuck to their guns. As she said, '. . . the results open up an entirely new field of research which unravels aspects of past human lifestyle far beyond basic biological reconstruction . . .'.

Since 1992 Dr Balabanova has tested a number of mummies from ancient Egypt. Nicotine showed up everywhere – for instance in three samples from Manchester Museum's collection and fourteen samples taken directly from an archaeological dig near Cairo.

The pathologist's results show that American tobacco was taken by ancient Egyptians as a matter of course. So who brought it to them?

I well remember one particular tomb on Crete. It's known as the Hagia Triada sarcophagus, a small limestone coffin of the Late Bronze Age (see second colour plate section). It is unlike most Cretan sarcophagi, in that it tells its own story. The tomb is decorated with a fascinating narrative scene, done on plaster. The painted frieze shows a sacred ceremony: a rite. It is intriguing that no similar painted sarcophagi have been found in excavations elsewhere on Crete. The joyful, pleasure-loving Minoans usually seem to have reserved fresco painting for the pleasure of the living, not of the dead. This may therefore be the tomb of someone who had travelled and was aware of customs elsewhere – especially in Egypt, with its tradition of painting tombs.

The puzzle is one that teases art historians greatly. The tomb shows, quite clearly, a man being carried off to his grave. As he is carried off, he is shown smoking a complicated pipe. (Some have interpreted it as a musical instrument, but I think this is unlikely.) It is clear to me at least that he has been either smoking tobacco or taking drugs, perhaps as part of a ritual, or more probably to dull the pain of whatever it was that ailed him.

I felt this was an enjoyable little coda to my revealing exploration of Egypt. Dr Svetlana Balabanova had proved that high-born Egyptians were fond of American tobacco; the tobacco beetle found in the ruins of Thera's buried city strongly suggests that the Minoans were the ones to provide it. In turn, they were given special status in Egypt. Perhaps some of them lived out most of their lives in a foreign land, adopting local customs, wearing local dress and generally 'going native'. Perhaps for the man on the Hagia Triada sarcophagus, this even meant adopting the Egyptians' special habit of drug-taking, before returning home to Crete to die.

The Hagia Triada sarcophagus has recently been redated to around 1370–1320 BC – to the time around the end of Egypt's 18th Dynasty.

14

RICH, EXOTIC LANDS

I am on my way to Mari, a hugely important former trading city on the western bank of the Euphrates. I am touring the eastern Mediterranean in search of the evidence – trade goods, vases, or even, should needs be, drugs – that will tell me more about the Minoans' influence in this region. Today this is an ordinary enough town on the Syrian border, named Tell Hariri. Once it was the site of King Zimri-Lim's exotic capital, an ancient Sumerian city finally destroyed by the powerful Hammurabi, king of the city-state of Babylon.

Mari was rich. Its position between Babylon, Ebla and Aleppo gave it control of key trade routes between the East and the West. The city collected taxes on all of the goods that travelled along the River Euphrates between Syria and Mesopotamia. The Old Assyrian traders paid taxes to the local rulers to try and protect their donkey caravans. In return, the rulers had to ensure that they would not be robbed along the way; if they failed they would give the merchants some recompense for the loss. Mari was also a key point on the land route that crossed the desert from northern Mesopotamia to southern Syria.

When Hammurabi attacked, Zimri-Lim disappeared from the record: we assume that he must have been killed. Over the centuries Mari itself was totally forgotten about and Zimri-Lim's rich and tremendous palaces were razed to the ground.

Nevertheless, at the beginning of the 2nd millennium BC Mari's magnificence was renowned throughout Mesopotamia. I am here to see a painting that shows the throne room at Mari, a scene which has remarkable parallels with the frescoes of Crete, including one in which a priest leads a bull to be sacrificed.

Mesopotamia's lost city was finally rediscovered when a large, headless statue was unearthed at Tell Hariri, on the west bank of the River Euphrates. The initial excavator of Mari, a French archaeologist named André Parrot, began by unearthing a large number of alabaster statues. He was convinced that what he was unearthing from the ground had strong connections to Crete. He particularly compared the painted stone imitations he'd found to those painted on dados from Knossos: perhaps the world's first examples of the technique of *trompe l'oeil*.[20] While a lot of the original frescoes are now in Paris at the Louvre, the whole southern façade of the 'Court of the Palms' of Zimri-Lim's 2.5-hectare (6-acre) palace, with its 260 rooms, has been reconstructed at the Deir ez-Zor Museum in Syria and is fascinating to see.[21]

Letters from the Mari archives give us countless examples of international intrigue and diplomacy: of diplomats, agents and spies who travelled extensively through the Bronze Age.

For now, I had decided to follow in King Zimri-Lim's footsteps and take a royal diplomat's view of the age. He had put a great emphasis on diplomacy, marrying off as many of his eight daughters as he could to regional rulers, in order to secure his networks of influence. Mari's official records, found in the 1930s and translated by French archaeologists, also give us the only detailed historical account in the world of a Middle Eastern diplomatic mission – or at least one that took place in the Bronze Age.

The records of Zimri-Lim's remarkable journey show how vital a role trade played in promoting the art and science of the time. Taking with him an enormous retinue of more than 4,000 men and coffers of gifts and tin ingots, the likable bon viveur travelled for six months. The records show that the king was an avid collector of Minoan art and pottery. Three months after setting forth, Zimri-Lim reached Ugarit on the Mediterranean coast, a scene of Minoan influence, where he stopped for a month, striking up a friendship with a local *danseuse* – not so different to today's Saudi princes in Beirut!

I will be following his trail and also tracking down many of the finds from Mari, which have ended up in the museums at Aleppo and Damascus.

Zimri-Lim's journey started at the beginning of the twelfth month of the Mari calendar, probably in mid April. As the king progressed north up the Euphrates he distributed gifts to the local rulers. Tin was the most sought-after commodity.

My own journey will last just a week but I expect to see some of the same sights (excluding the dancers …) as in Zimri-Lim's day, nearly 4,000 years ago. On the way, I re-read Jack M. Sasson's account of that Bronze Age world:[22]

> The terms 'global' and 'multicultural' are often applied to con-temporary society, which has just stepped out of the 2nd millennium AD. Remarkably, such concepts were relevant as well to the 2nd millennium BC, when building upon the momentous developments of prior millennia – the origin of cities and the invention of writing. An expanding social elite required bronze and demanded exotic luxury goods from distant lands. These needs fostered the creation of an era of intense foreign contacts, with new technological break-throughs such as the invention of glass and a revolution in travel.

This global culture was astounding. Sasson's research reveals a world in which kings exchanged gifts of extraordinary beauty and elaboration; salt cellars in the shape of lions and calves; drinking bottles in the shape of a horse, inlaid with gold eagles and lapis lazuli; even, as in the Uluburun wreck, a golden chalice fit for a hero.

Mari's official records show just how full and sophisticated a trader's life could be. As well as trading in basic goods – livestock, grain, oil and wine and raw materials such as wool, leather, wood, reeds, semi-precious stones and metal – they dealt in exotics – models of Cretan ships, desert truffles, rare wild animals including bears, elephants and wild cats. Of equal importance to the development of civilisation, countries of the eastern Mediterranean and Mesopotamia exchanged intellectual ideas: concepts in philosophy, science and religion.

In this cultured environment, anyone who had a valued, specialist skill could move freely from place to place – astronomers, physicians, translators, gymnasts, cooks and seamstresses. Travelling artists were especially prized. Musicians came from Qatna, Aleppo and Carchemish.

Cowries from the Maldives were used for currency. How did these cowries get to Mari when the Maldives are deep in the Indian Ocean? Jewellery was made from lapis lazuli from Afghanistan and turquoise, jadeite, cornelian and quartz from India. By 2680 BC cedar was imported from the Lebanon and glass beads in their thousands from Indian rivers. Perhaps the best-known treasures from Mari are the proud bronze lions, now in the Louvre.

Most of all, the Minoans drove the process of international development – they had the ships and they brought the essential raw materials upon which the crack pace of Bronze Age civilisation relied – copper and tin.

Travel across Mesopotamia was for the most part by river. The River Euphrates Zimri-Lim would have known teemed with boats, barges and rafts, perhaps even more than it does today. Levantine cities put on concerts with varied and colourful programmes – men belched fire and swallowed swords; jugglers, wrestlers, and acrobats performed for the public; as did actors and actresses in masques and plays. New ideas and inventions travelled up and down this vital causeway. I settled back to enjoy the ride.

Next stop, Beirut. Before the civil war of the 1970s and 1980s, the Beirut I knew was an enchanting city, with its backdrop of cedar-clad mountains. It used to be a haven for Saudi businessmen. Here they could escape the pressure of commercial life in Saudi Arabia: they could gamble in lush casinos, eat pigeon *libanaise* in the glitzy restaurants which lined the corniche and have their pick of the local dancing girls for a night. You could ski on the placid seas and on the mountain snow on one and the same day.

Today's Beirut, sadly, remains partly ruined. Happy memories have gone. I cannot get out of the city fast enough, so I haggle with minibus drivers to take me across the mountains to Damascus. The price is usually halved by evening. That way the driver can return with a van load of vegetables from the Bekaa valley, in time to sell at dawn in Beirut's markets.

Like the Lebanese, the Minoans were master traders. I thought about that principle – if you travel one way carrying human cargo, when you return you maximise your assets by bringing something else back with you. Perhaps the Minoans did the same.

At the Syrian border, Ahmad, our driver, plonked our passports in front of the Syrian customs officer. Fat, agitated and energetic, Ahmad ate continuously. His side pocket bulged with seeds, which he nibbled.

There was something odd about this minibus. Perhaps that explained why the customs officer is not being obliging enough. Ahmad pulls out a wodge of Syrian pound notes which the customs officer pockets without a trace of emotion, or thanks or receipt; now we can set off again. Once in no man's land Ahmad stops to collect eight full black plastic bags which he hides in the spare tyre bay. I hope these are not drugs. If not, we should have a clear run.

Damascus claims to be the oldest city in the world – a title also fought over by Samarkand, Bukhara, Aleppo and Cairo, among others. From late November to early March, the River Barada carries rainwater down to the plain, creating a large, rich oasis named 'the Ghouta'. The city not only has a beautiful climate, rich soil and abundant water but was at the crossroads of ancient trade routes (see map). One route led north from Egypt up the fertile crescent through Damascus and on to Mesopotamia. Coming from the opposite direction, a merchant from the East landing from India in the Euphrates estuary could travel upriver through Mesopotamia then turn south through fertile land, by-passing the mountains all the way to Egypt. I intended to explore the museums, to get a snapshot of the trade and civilisation of 5,000 years ago.

For thousands of years Damascus has been famed for her craftsmen, masters of inlay on wood – pearl inlaid on rose, walnut or apricot – her women and her precious damasks. Yet Damascus Museum – possessing probably the finest collection of sumptuous Mesopotamian art in the world, stretching back over five millennia – is still a disappointment, with little information to be had and still less scope for research. The few guidebooks are sycophantic to the point of hilarity. One introduction reads: 'To President Al Assad, whose march of correctionism is an inspiration and a stimulus'!

Although I had found plenty to interest me, I hoped to find more accurate information about Bronze Age artefacts at Aleppo. At the bus station, a stunning Syrian girl seems to be expecting me. 'Sit down – I will get you a ticket.' She takes my money and touts around the buses to find one which will take me for 200 Syrian pounds, then pockets the change. Soon we are rolling northeast in a comfortable Mercedes 403. The road is on a dividing line. To the west the mountains and the sea, to the east a lush plain watered by the Barada River.

Seven hours out of Damascus the bus pulls up in central Aleppo opposite Baron Street. There are four hours of daylight left – just enough to visit the souk. I dump my bags in Baron's Hotel and hurry to the marketplace, a vast space that still, today, eventually leads to a copper market. What an amazing experience – little has changed for 5,000 years. It is covered by great stone archways for some 18 miles (30 kilometres). The first line of stalls is for butchers who sell sheep's testicles (at a cost of 90p each) and nothing else. They are huge, each the size of a squashed orange. The butchers, I notice, all specialise in different parts of the animal: the first in the pancreas, then the liver . . . a group of the next few stalls are selling hooves and tails. I have walked for a mile and have only seen bits of sheep! I explored the place solidly for four hours: there must have been 50 miles (80 kilometres) of vaults and stalls. The organised commercial districts are known as Khans. At Khan al Nahasin (the Khan of the coppersmiths) is Aleppo's oldest continuously inhabited house, which has been kept almost exactly as it was four centuries ago. It was once lived in by a man called Adolphe Poche, whose parentage was both Venetian and Belgian. Poche was born in this house and yet became Belgian consul to Syria in 1937. Appropriate for Aleppo, I suppose: for thousands of years it's been one of the most important trading cities in the world. Only a few hundred kilometres away from the Mediterranean, this ancient city is the meeting point of the two oldest land trade routes known to man.

Baron's Hotel was founded in 1909 as a lodge for Mr Baron to relax in after duck shooting. Little has been altered since then; perhaps not even the bedclothes. It is a delight to be here. Slightly foxed posters in the rather *louche* bar – a mix of 1970s high stools and what looks like 1940s everything else – advertise the inaugural Orient Express train journey to Aleppo. Mounted above the bar is Lawrence of Arabia's bill – extravagantly high, at £72.09. A young female French archaeologist and I are the only guests.

The crime writer Agatha Christie and her archaeologist husband Professor Mallowan stayed here for months. Professor Mallowan had left the hotel a hand-drawn map. Stretching away in an arc to the east of Aleppo, it showed the mass of extraordinary archaeological sites of the middle and upper Euphrates. There are between fifty and a hundred sites, going back 5,000 years. In the very early dawning of the Bronze Age this area was the most heavily populated in the world.

Breakfast is at dawn. The French archaeologist, a dark young woman with a pinched face, offers me her boiled egg and cheese provided I do not talk to her.

Then we set off into the rising sun in Ahmad's battered minivan. The land is flat as a pancake; the horizon flirts with infinity. On the outskirts of Aleppo the figs and apricots are ripening. Further out come poplar plantations, then the endless vista of rich, cultivated land. Red earth under plough, bilious green rice, thin winter barley and dark ochre stubble where the last of the cotton has been harvested.

Cotton stalks are piled high beside the farmhouses for winter kindling. Flocks of fat-tailed sheep and slim black turkeys scavenge the fields. Each house has its own pigeon loft. Skeins of duck, high in the sky, migrate south. No one is shooting them now. I suspect that there is much more to discover about Aleppo, but I am happy with what I found in the museum, including some cuneiform slates about Mari that were originally discovered by Max Mallowan. Mounds of quinces and watermelons lie beside the road.

15

PROUD NINEVEH

Nineveh, at the centre of what was Babylonia, is less than a day's travel and my final destination before I return to Beirut. I was here, again following in the wake of the Minoans, because this great city was once the epicentre of world knowledge. It was here that the Minoans could have pursued their understanding of the heavens to sacred levels. This time, I was on the trail of some sacred omens; omens written down by trusted Babylonian priests during the Old Babylonian period (1950–1651 BC). This would have been hard-won information, collected for generations and then written on clay.

The ancient mounds and ruins lie at the crossing of the Tigris and the Khosr rivers, near the modern-day Iraqi town of Mosul. The 'exceeding great city', as it is called in the biblical Book of Jonah, lay on the eastern bank of the Tigris in what was ancient Assyria. It is now one immense area of mounds and broken walls, overlaid in parts by rackety modern suburbs.

In its calm and quiet, Nineveh's atmosphere must have once been like what we would today call a university town, like Oxford, Salamanca or Bologna. And yet here, in what's now the Iraqi desert, all that remains are humps of rubble and mounds of bare earth. I was reminded of the biblical prophecy against 'proud Nineveh': 'And He will stretch out His hand against the north and destroy Assyria, And He will make Nineveh a desolation, Parched like the wilderness.'[23] That's pretty much what seems to have happened.

But it was here that the great Assyrian king Ashurbanipal, or Aššurbanipal, had his palace and a library of world renown. Under his rule the Assyrian kingdom stretched as far as the Gaza Strip to the west, Armenia to the north (towards the Black Sea), east towards the Caspian Sea and the Persian Gulf to the south.

Aššurbanipal reigned much later than the Minoans' time, in the 7th century BC. Assyria's genius had been built on the foundations of extreme military might, the imposition of ruthless discipline on its people and the exercise of extreme brutality over those it conquered. The Assyrians seized Babylon in the 8th century BC. Yet this contact at least seems to have been a civilising one, inspiring the Assyrians to educate themselves. Aššurbanipal was the ruler who finally managed the impossible: uniting the two traditions of Mesopotamia – war and words – within one culture. His significance for me was that, far-sightedly, he had put together a collection of much older astronomical and scientific texts. This was sacred knowledge, all of which he had ordered to be collected together from all over Mesopotamia, not least from the already ancient cities of Babylon, Uruk, and Nippur. His collection began with the work of the Sumerians, to whom we owe so much of our own modern-day culture, including the division of time into twelve- and six-hour blocks. Aššurbanipal's library was still in use when Alexander the Great defeated the Achaemenid King Darius III and conquered Mesopotamia.

When it was unearthed the library painted a vivid picture of a violent past; it also held a version of the biblical story of the great flood. Aššurbanipal's collection also proves that as far back in history as you can look, humanity has been obsessed with the heavens. The seasons ruled people's lives. Farmers calculated what work they had to do, and when, according to which constellations were rising and setting at dawn. At that time of mystery and wonder, celestial movements in the sky must have seemed like the jousting of the gods. The constellations were seen as miraculous things; they still inspire wonder today, thousands of years after Homer's hero Odysseus made his slow way home by the stars:

> Sleep did not fall upon his eyelids
> as he watched the constellations – the Pleiades,
> the late-setting Boötes, and the Great Bear,

which men call the Wain, always turning in one place,
keeping watch over Orion – the only star
that never takes a bath in Ocean. Calypso,
the lovely goddess, had told him to keep this star
on his left as he moved across the sea.[24]

I was here because I had a problem. My theory that Minoan ships could cross the Atlantic depended on one thing: navigation.

It is relatively easy to find latitude at sea. One way is to calculate the angle of the sun in relation to the horizon, at dead noon. It can be done using a very basic quadrant. With three pieces of wood and a little luck (no cloud or rain!) you can calculate your latitude to a fair degree of accuracy. You can also use the night sky and you can even use the simplest equipment of all, your own arms, to do it. In the northern hemisphere, all you need to do is point to the North Star, Polaris, and extend your other arm to the horizon. If the angle is 30 degrees, you are at 30 degrees north. At the equator – zero degrees latitude – Polaris appears to be on the horizon line.

Longitude is a very different matter. It is much, much harder to calculate. It was the greatest problem for navigators in Europe up until the 18th century. Yet from my initial investigations into ancient records I had a strong feeling that the Babylonians had found a way of establishing longitude as far back as 1300 BC. And that their Minoan trading partners shared that knowledge.

Could the Minoans truly have navigated well enough to be the world traders I thought they'd been? If so, how had they done it?

||||||||||||||||||

In the West, we think it was Copernicus who first realised that the earth and the planets orbited around the sun. The truth, I'd discovered, could hardly be more different. It is clear that the Babylonians had realised this. How had they come upon such remarkable levels of knowledge?

The answer lies in their extraordinary dedication to stargazing. To begin with, this was nothing to do with navigation. They believed that the gods had created the movements of the planets to help people on earth tell the future. The stars were used like a horoscope, to predict future happenings – and to try to avert catastrophes. One such prediction ran:

> When in the month of Ajaru, during the evening watch, the moon eclipses, the king will die. The sons of the king will vie for the throne of their father, but will not sit on it.[25]

Apparently, just before a disaster like this was predicted to happen, the king would temporarily abdicate his throne. A substitute then took the crown. If the prediction was death, the unfortunate replacement would be killed. What's known as a self-fulfilling prophecy ... or having your cake and not eating it.

So the Enuma Anu Enlil tablets, preserved for posterity by Aššurbanipal at Nineveh, are full of astronomical events that successions of Babylonian peoples and their kings had been charting, documenting and collecting for many generations. Astronomers worked for centuries, detailing exactly which stars rose on a particular day, the angle they rose at, at what time – and the distance between the stars. For instance, they knew that a different star rose on the eastern horizon each sunset over a time span of four years and they realised that after that the cycle repeated itself.

Some of the tablets are missing and some are difficult to decipher. Yet many describe clearly the timings of moonrise and moonset, the rising and setting of the planets and the patterns of both solar and lunar eclipses. Tablets 1 to 22 (dating from around 1646 BC) describe the moon's movements; tablets 23–36 the sun's eclipses, coronas and parhelia; tablets 50–70 the planetary positions; and tablet 63 shows the movement of Venus.

By the time of the final, mathematical phase of Babylonian astronomy, so much data had been collected that scribes were able to calculate what was about to happen in both the day and the night sky simply by looking back in their records. Ephemeris tables (using both the sun and the stars) and sidereal tables (solely the stars) are essential to navigation. They show the day-by-day positions by sign and degree of the sun, the moon, Mercury, Venus, Mars, Jupiter and Saturn. The *Oxford English Dictionary* defines them as 'a table showing the predicted (or, rarely, the observed) positions of a heavenly body for every day during a given period'. In short, an almanac.

Supposing the Minoans had relied on their trusted trading partners for reliable astronomical tables that would help them with their astronavigation, especially for determining latitude at sea with accuracy. Perhaps they'd taken that precious knowledge and worked out ways to produce their own star charts, for nearer home. I had to find out.

I read and read: as a submarine navigator I'd spent many years of my life calculating both latitude and longitude, sometimes by using meridian passages of the moon. So I could anticipate some of the practical problems that the Bronze Age navigator must have faced. There are two essentials: having a fixed point, or observatory, as your reference point, and – and this is the killer – knowing the exact time of day.

In navigation, time translates as distance. Most sailors know the adage: 'Longitude west, Greenwich time best. Longitude east, Greenwich time least.' In other words, travel east and you are ahead of Greenwich Mean Time. Go west, and you are behind it – or in other words, later than GMT. Thus, if you live near Greenwich, London, never ring a friend in Greenwich, New York State at nine in the morning. The chances are you'll get someone with a bear for a head.

Unfortunately, when navigating, you cannot get this wrong. While miscalculating a minute of latitude could put you out by a negligible 1.15 miles (2 kilometres), degrees of longitude vary in size, getting smaller towards the poles, where the meridians converge. If you guessed the time when trying to find your longitude at sea, you could easily be 'out' by 1,000 miles (1,600 kilometres) without knowing it.

Nowadays, it only takes a few seconds to download pre-calculated ephemeris tables on to your computer. However, the moon is nothing if not changeable. Calculating these lunar ephemerides is so complex that in the 18th century – all knowledge of Babylon's tables having been wiped out by history – vast riches lay in the path of those who could find a new method of doing so. If my understanding of tables 20 and 21 of the Enuma Anu Enlil was correct, the extraordinary Babylonians were able to predict eclipses of the moon over its entire, 18.61-year cycle. Whether or not this enabled them to calculate longitude I was not sure. It is beyond strange to think that man already had this sacred knowledge 1,500 years before the ancient Greeks and more than three millennia before Copernicus or Galileo.

16

THE KEY TO INDIA?

Having arrived back in Beirut, I kept thinking back to Mari, on the middle Euphrates, to the northeast of Damascus. Diplomatic missions crossed frontiers and the Minoans traded here extensively. Intriguingly, by 2680 BC Indian glass beads in their thousands and cowries from the Maldives were used for currency.

Where were the Minoans leading me now? So many of the items found at Mari and now housed in Aleppo's museum seemed to have had their origins in India. How did all those cowries and beads I'd seen in the museums get to Mari – when the Maldives are deep in the Indian Ocean? From their base in Tell el Dab'a, could the Minoans have got to India? It took some stretch of the imagination to dare to think that. I had travelled all this way: it had taken me weeks and months of planning to do so. Imagine the challenges for a traveller during the Bronze Age.

My mind turned back to the Indian elephant tusk I'd seen at Heraklion on Crete: excavators had found it in the ancient town of Zakros. There was also a similar one found packed into the hold in the Uluburun wreck. From the Yemen to India is some four weeks' sailing on the southwest monsoon, which starts in June. Let us push on to India, in the wake of the ghosts of the Minoan fleet. Perhaps there is evidence that they did in fact reach the subcontinent somehow, travelling via the land mass of Egypt, just as the Egyptians had evidently reached the fabled Land of Punt?

To my astonishment I discovered that a considerable number of hoards of bronze goods, deliberately hidden underground, have been found over much of northern India. There have been 129 of them to date, most frequently found near the Ganges – in the Jumna catchment area. Most of these sites have been discovered by local farmers ploughing their land rather than by controlled excavation, so it has not been easy to date them. However on a number of sites a distinctive ochre-coloured pottery has been found which has been much easier to date – to the 2nd millennium BC, once again, which according to all the evidence was the primary era of Cretan ascendancy and contemporaneous with the Minoan palace at Tell el Dab'a.

Typically these Indian Bronze Age hoards consist of harpoons, swords, rings, chisels and axes, including double axes similar to those of Minoan Crete. The hoards have several noteworthy features. Firstly, they rarely include the implements and tools used by Indian village people – such as the knives, digging tools and arrowheads that you imagine would have been useful in the daily life of Indians as it was in the 2nd or 3rd millennium BC.

Secondly the blades or cutting edges of the tools are seldom worn or chipped – they do not appear to have been actually used. They are more like samples, or stores, carried for sale.

Who were these travelling salesmen of the Bronze Age? A possible answer is Minoan traders in fleets operating from their base in Egypt. It could, of course, be a series of coincidences that the Indian copper hoards are of the same era (the 2nd millennium BC). It could be another coincidence that they contain unused double-headed axes, of exactly the same design as the distinctive Minoan ones and that these implements were foreign to India (in the sense that they were not used by the local people as tools).

The possibility of coincidence would be greatly reduced if there was evidence of trade between India, Egypt, and the Minoans in the 2nd millennium BC. I was going to have to go back to my research sources: there was a lot I had yet to understand. The bit was firmly between my teeth.

THE SEARCH FOR THE RED SEA–NILE CANAL

... And if they did reach India, then how? The key lies in that rich strategic relationship that Crete had with Egypt. Using Tell el Dab'a as a base meant the Minoans could load their ships with dates, fresh vegetables and salt fish ready for expeditions to India and the East. Queen Hatshepsut's well-documented expedition to Punt in the summer of 1493 BC was prepared for in exactly this way. She had sent a fleet of five ships with thirty rowers each from Kosseir, on the Red Sea. Where would the Minoans go from Tell el Dab'a? The answer, I suspected, would be found near the Red Sea–Nile canal.

In 1998 a group of stone megaliths was found at the coastal plain of Tihamah, Yemen (see map). The site was investigated by the Royal Ontario Museum, Canada and the Yemen Government. Beneath the standing stones they found a hoard of copper and bronze artefacts and tools, dated between 2400 BC and 1800 BC. As Edward Keall of the Royal Ontario Museum said at the time:

> We didn't know what was keeping people in this terribly marginal desert area ... Was it a natural resource or a strategic position that prompted these people to invest such effort in creating these remarkable monuments? (www.archaeology.org)

Why else, I thought to myself, than because of the Red Sea–Nile canal? 'King Scorpion' is said to have been the very first Egyptian canal builder. His superb macehead is now in Oxford's Ashmolean Museum, like many fascinating artefacts from the Bronze Age, including the controversial fresco fragments from Alalakh. After an extraordinary refit the museum has emerged, blinking, as it were, into the new light. It was once something of a labyrinth of its own, where it was hard to find all but its most famous curiosities, such as Lawrence of Arabia's cloak. The Minoan collections are fascinating: they include a six-tentacled octopus storage jar of around 1400 BC, a large decorated *pithos* from the storerooms of Knossos and weapons from the so-called 'warrior graves'.

Once I discovered that Arthur Evans had worked there, I soon began to explore the museum for inspiration as well as information. In Evans' archive is a photograph of a pillar being excavated in the east pillar crypt at Knossos. The symbol of the *labrys* – the double axe – is marked on every surface.

Meanwhile 'King Scorpion's' huge pear-shaped macehead shows a threatening scorpion hovering in the air. Wearing the tall white crown of Upper Egypt, the king stands on the bank of the Red Sea–Nile canal, with a digging implement in his hands. Down below, the king's workmen are seen putting the final touches to the canal banks. The limestone mace dates from the 4th millennium BC.

We know little about this pharaoh save that he conquered part of the Delta. The first king after 'Scorpion' whom we can date with reasonable accuracy is King Menes, who lived around 3000 BC in his palace at Memphis. According to Herodotus he dammed the Nile some 12 miles (19 kilometres) south of Memphis and directed the waters to form a new lake linked to the Nile by a canal. In the 6th Dynasty (c.2300–2180 BC) Pepi I planned a canal through the first cataract – to tame it. The canal was cut by Uni, the governor of Upper Egypt. Sometime during the Middle Kingdom (2040–1640) a canal was dug between the Red Sea and the eastern branch of the Nile at the Delta.[26] Using captured enemies as slave labour, Egypt set off on an orgy of water-channel-making, so much so that the face of the nation was completely changed.

... All Egypt is level; yet from this time onwards it has been unfit for horse or wheeled traffic because of the innumerable canals running in all directions, cutting the country into small segments. It was the King's desire to supply waters to the towns which lay inland at some distance from the River, for previously when the level of the Nile fell, the people went short and drank brackish water from the wells. It was this King also who divided the land into lots and gave everyone a square piece of equal size and from the produce extracted an annual tax ... Perhaps this was the way in which geometry was invented ...[27]

Herodotus says Egyptian priests informed him that at one time the Red Sea and the Mediterranean were connected. Thousands of years later, Napoleon carried out a cadastral (land boundaries) survey after his conquest of Egypt in 1798. His maps of the Delta and the Red Sea–Nile canal may be viewed in the Louvre. The Red Sea–Nile canal is also shown on a British cadastral survey of 1882. By comparing this 1882 map with Napoleon's maps and Google Earth, even today one can see the route of the ancient waterway. After heavy rain in particular, you can trace its course on satellite photos as it passes under Ramses II Street, emerges in northeast Cairo and then heads toward Zagazig, in the eastern part of the Nile Delta. You can follow its faint outline all the way to the Red Sea.

|||||||||||||||||||||

Assuming for the moment that the Minoans did use the Red Sea–Nile canal, quite possibly in company with Egyptian sailors or ships, they would have entered the northern Red Sea – the adventurer's route to the heady lands of Punt and India. If the Minoans had used the canal frequently, I reasoned, there could still be some concrete evidence left. I wondered if I could uncover any of the traces left behind, in the form of Bronze Age ports or buildings on the land which borders the Red Sea – either in Egypt, Arabia or the Yemen.

The hoards of bronze near the stone megaliths of Tihamah speak of trade. A photograph of axe heads in the hoard is shown in the first colour plate section. Not only that, but the megaliths themselves also tell their own story, a story I shall return to later on in this book.

The important thing is this. From the Yemen, reputedly the Land of Punt, to India is some four weeks' sailing on the southwest monsoon, which starts in June.

17

INDIAN OCEAN TRADE
IN THE BRONZE AGE

Sailing in the Indian Ocean is determined by the monsoon wind, which is caused by the difference in temperature between the massive Himalayan Plateau and the sea (see map). In summer the Asian land mass becomes hotter than the ocean, sucking winds and water vapour off the sea. In April the southwest monsoon is heralded by westerly winds in the Indian Ocean. By May the southwest monsoon hits Indochina, to reach its peak and constancy in July, by which time winds reach 30 knots in the South China Sea. By then, India is inundated with monsoon rain. During September the temperature drops and by November, when the Himalayas have become bitterly cold, air is drawn off the mountains by the warmer seas.

The northeast monsoon starts in late December, after which the wind gradually abates until April, when the cycle begins again. Sailing ships voyaging between Egypt, Africa, India and China would have had to take advantage of the monsoons in order to sail before the wind, returning on the next monsoon to their respective countries. They awaited the change in a sheltered harbour. Hence the need for capacious ports around the Indian Ocean, where goods could be stored from one monsoon season to the next.

Monsoons are so predictable – and so important – that they were later incorporated into calendars, which illustrated the highly synchronised system of regular shipping between Egypt, East Africa, India and the Gulf. For example one such calendar has this for day 68 (March 16): 'End of sailing of Indian ships from India to Aden: no one sets sail after this day' and 'on day 100 (April 15) the last fleet from India was scheduled to arrive in Aden ... on Aug 14 (day 220) the last ship from Egypt arrived in Aden. Six days later ships from Sri Lanka and Coromandel set out on their voyage home.'[28] The last departure from Aden, powered by the monsoon, was on day 250 (September 13).

In short, ships sailing from Egypt for India would be carried by the southwest monsoon which ends in September. They could then trade in India until it was time for the northeast monsoon starting in December, which would carry them back home to Egypt and, via the Red Sea–Nile canal, to the Mediterranean. They had a free ride each way. The west coast of India has many great rivers carrying melted snow down from the mountains. Their estuaries provide the opportunity: wonderful ports could be built in the shelter of most of them, from which to export the riches of India. Marcella and I decided to take advantage of an invitation to speak at a naval academy event. The coasts of India beckoned: and who were we to gainsay that?

I reasoned that there should be Bronze Age ports from Karachi in the north of India all the way down the coast to Kerala, in the south. As I began my research, I realised that three ports – Lothal, Cambay and Muziris – were important in the Bronze Age, specialising in exports.

LOTHAL – AN INDIAN BRONZE AGE PORT

In a series of articles published between 1955 and 1962, Professor S. R. Rao, an Indian marine archaeologist, describes his excavation of a port at Lothal, inland from the Gulf of Cambay in northwest India. Lothal was then much closer to the sea than today (see map).

In those seven years, Professor Rao and his team unearthed channels and locks leading from the river to a large rectangular port area. The dockyard itself was lined with well-made bricks and was designed to control the flow of water through a sluice from the river into the dockyard. Lothal had the world's first lock system.

The town that surrounded the dockyard was built between 2500

BC and 1900 BC. The dockyard itself provided sheltered mooring in an enclosed harbour measuring 214 × 36 metres (702 by 118 feet) – absolutely enormous for 4,500 years ago. The entire settlement was divided into two parts: a citadel or acropolis for the ruler and wealthy merchants and a lower town for the workers. Acropolis houses were built on 3-metre-high (10 foot) brick platforms and were provided with running hot and cold water, a well for drinking water and a sewage system that was designed for flushing out and for solid waste to be removed. A large warehouse which was situated at the southwest corner of the Acropolis was again raised on 3-metre stands. Goods were protected from flood or theft by being raised from the ground, by wooden walls on four sides and by a wooden roof.

The surrounding land was well watered and produced fine Indian cotton and plentiful rice. The sea coast provided shellfish and the river had beads in profusion.

Thus, with an abundance of fresh water and with a river leading to the sea, Lothal became the most important port in India – and from the point of view of archaeological finds one of the richest sites in the subcontinent. These finds are now exhibited at Lothal in a modern archaeological museum that was established in 1976. To summarise the official description:

> The museum has three galleries. In the front gallery, a canvas depicts an artist's impression of how the town was laid out. There are also introductory maps and descriptions to explain the importance of Lothal. The left hand gallery displays showcases with beads, terracotta ornaments, seals, shells, ivory, copper and bronze objects, bronze tools and pottery. The right hand gallery has games, and human figurines, weights, painted pottery, burial and ritual objects and a scale model of the whole site. (www.indianetzone.com)

The famous Indian beads found in local rivers are of cornelian, agate, amethyst, onyx, semi- precious stones and faience. Tiny micro beads can be seen through magnifying glasses.

The seals are engraved on steatite with Indian writing and animal figurines on the face. Shells are made up into bangles, necklaces, games and musical instruments.

Copper ingots of 99.8 per cent purity were imported, as was tin. These were smelted to make a wide array of weapons, tools and cooking implements. Pottery, including huge *pithoi* for storage, came in all shapes and sizes. There were games made of bone, shell and ivory as well as clay figurines representing subjects such as a gorilla, or humans. The gold work was extremely fine, with minute golden balls that require a microscope to view them properly. The people who had made the artefacts had a standardised weight system, using weights made of cornelian, jasper, agate and ivory.

The similarities between these archaeological artefacts excavated from Lothal and those found in Minoan sites or Minoan wrecks are striking – even astounding. The same can be said for the layout and construction of the town.

A comparison between forty-six artefacts excavated at Lothal with those at Minoan sites or in Minoan wrecks of the same era – the 3rd and 2nd millennia BC – discloses that all forty-six are very similar, or identical. Could these artefacts have developed independently? Did the Minoan and Indian civilisations develop at the same time, needing the same goods? I believe this argument breaks down for three principal reasons. Firstly, Lothal imported copper and bronze. For reasons which will be considered in more detail later, the copper ingot found at Lothal was of over 99.8 per cent purity. The only mines which produced copper of that purity in 2500 BC were the mines of Isle Royale and Lake Superior. Ships must have brought that copper, crossing the Atlantic to do so. Minoans had ships capable of such journeys. Secondly, items unique to India – elephant tusks and Indian beads found in Indian rivers – have been found in Minoan palaces and wrecked ships (Uluburun). So, I would argue that Minoan ships must have sailed to India to collect those items. Thirdly, it's worth noting the scale of the coincidence – if there were a dozen excavated artefacts one could just about attribute it to chance, but 46 ?! You can make up your own mind by visiting the gallery pages on our website.

What is I think possible, even likely, is that there was a substantial exchange of goods and ideas. Minoan ships took copper and tin to India and returned with ivory and cotton and perhaps many Indian ideas about town planning, civic engineering and astronomy. Sooner or later, Indian shipowners would have wanted their vessels to accompany Minoan ships across the world, to collect valuable goods for themselves.

It is time to leave Lothal and travel southwards once more down India's beautiful seaboard. We had intended to visit Cambay, north of Bombay, knowing Cambay had been a great international port in the Bronze Age. However we learned that the town is now beneath the sea, due to a shift in the continental plates. So we must push on further south knowing little of the ports of southern India.

How can we narrow our search to find ports which flourished in the 2nd millennium BC? I've prepared a plan, which involves starting with the accounts of those celebrated authors who had described trade in prehistoric times. Professor A. Sreedhara Menon, in *A Survey of Kerala History*, has provided a very useful summary. He describes classical writers giving vivid accounts of the thriving spice trade between the Kerala coast and the Roman Empire through the ports of Muziris (South India), Tyndis and Barace – the classical writers being the Greek ambassador Megasthenes (4th century BC), the anonymous author of *The Periplus of the Erythraean Sea* (1st century AD) and Ptolemy (2nd century AD). And the Peutingerian Table, a set of maps dating from about AD 226 that is reputed to have been copied from fresco paintings in Rome, is said to show a temple of Augustus near Muziris and a Roman army being stationed in Muziris for the protection of the Roman spice trade with India. The location of Muziris will be discussed later.

To match these Roman and Greek accounts, Professor Menon cites Sanskrit works. The *Mahabharata* mentions the king of Kerala providing provisions; the 4th century BC *Arthashastra* mentions the River Periyar as one of the rivers of Kerala where pearls can be found. The *Puranas* also mention Kerala. Apart from Sanskrit (language of North India), Tamil writings in the language of South

India are also important sources of information: ancient Tamil literature is replete with references to the land of Kerala, its rulers and its people and its well-developed civilisation. The *Patittupattu* (Ten Decads) is an anthology of 100 poems which reconstruct the history of ancient Kerala and her trade.

To these classical, Tamil and Sanskrit accounts, we can add those of the Chinese: Hiuen Tsang in the 7th century AD; Wang da Yuan in *Descriptions of the Barbarians of the Isles* (1330–49); and Ma Huan, who accompanied Admiral Zheng and describes Cochin and Calicut with great verve.

Arab writers al-Idrisi (AD 1154) and Yaqut al-Hamawi (1189–1229) give descriptions of Kerala's coastal towns and their trade, Al-Kazwini (1236–1275) provides information about Quilon and Dimishqi (AD 1325) writes about the Malabar coast. Also, the accounts of Ibn Battuta describe his six visits to Calicut and the pepper trade operating out of the port of Quilon, where there were huge Chinese junks.

Finally we have the early European travellers who describe a very old spice trade between Kerala and the Arab, Mediterranean and Chinese worlds – Friar Odoric of Pordenone who reached Quilon in 1322 *en route* for China; Friar Jordanus of Severic who came to Quilon in 1324; the papal legate John de Marignolli of Florence who lived in Quilon for a year; and Nicolò da Conti (1420s–30s) with his description of the ginger, pepper and cinnamon trade of Quilon and the jack fruit and mango trees along the coast. The Persian ambassador Abdul al-Razzak (1442) testifies to the rich Malabar trade with the Arab world, as does the Russian Athanasius Nikitin in 1468–74.

In short, accounts from many sources stretching back thousands of years testify to the fact that the Malabar coast of Kerala and her ports of Calicut, Cochin, Quilon and Muziris traded valuable spices with the Arab world, the Mediterranean, Africa and China.

LOCATING MUZIRIS

The port whose name crops up time and again is Muziris. Muziris was an important spice port long before Roman times. We therefore intended to locate Muziris and, having done so, to see whether there is evidence that Minoan ships traded there.

Muziris is likely to have been a natural harbour near where the most valuable spices, that is pepper and cardamom, grew. We can narrow our search by locating the best growing areas for pepper and cardamom.

Southern India has a peculiar geography. It is near the equator and therefore will have an equatorial climate, but this is modified by a range of mountains called the Western Ghats, which run north–south, parallel to the coast for some 995 miles (1,600 kilometres). Their average height is 900 metres (2,950 feet). The Ghats are punctuated by a number of wide valleys which allow monsoon winds to funnel through. The result is that there are three seasons: summer, rainy season and winter. The coast is hot and wet but it is cool and pleasant in the hills with light sea breezes in the foothills. Pre-monsoon rains called 'mango showers' are beneficial to coffee and mangoes. The southwest monsoon arrives at the end of May with 2 to 4 metres (6.5 to 13 feet) of rain all along the coast. This high rainfall, high humidity and a long wet season has given rise to dense, evergreen luxuriant vegetation ideal for palm trees – Kerala means *kera* (palm) and *la* (land). The cool, moist hill slopes of the Ghats provide ideal conditions for tea, coffee and spices. Kerala is the world's largest producer of cardamom, the most valuable spice, today costing four times as much as black pepper. The coast of Kerala, from Calicut in the north to Quilon in the south, is characterised by a number of lagoons called backwaters, providing a series of internal waterways – canals protected from the sea by sandbanks, resulting in wonderful natural harbours. Moreover, into those protected harbours flow no less than forty-four rivers, which rise in the Western Ghats. In short, along this stretch of coast, named 'The Malabar Coast' by the British Raj, are endless fine protected harbours. Rivers lead explorers quickly to the foothills of the Western Ghats, where pepper and cardamom flourish. The elusive Muziris, confusingly known by a number of ancient names including Shinkli, is therefore likely to have been situated along this 620 mile (1,000 kilometres) stretch of coast (see map). This preliminary conclusion appears to be borne out by a huge hoard of Roman coins found inland near Palghat in central Kerala.

THE LOCATION OF MUZIRIS

In 2006 the Kerala Council for Historical Research (KCHR) found a prehistoric Bronze Age site at Pattanam, 25 miles (40 kilometres) north of Kochi (Cochin), near the estuary of Kerala's largest river, the Periyar. The research team, headed by its director, Dr P.J. Cherian, started excavations near to the position of previous finds of Roman amphorae. An earlier team had found Roman coins, a bead chain and Roman artefacts nearby. To quote extracts from KCHR:

... The third season (2009) archaeological excavations at Pattanam reiterates the assumptions that Pattanam might be the oldest port site with extensive evidence for Roman contacts on the Indian Ocean rim or beyond ...

... The initial inference from the field is that the majority of the samples [of pottery] are of the campanian type of south Italian origin with volcanic elements. Greek sources such Kos and Rhodes and Egyptian and Mesopotamian amphora sherds were also found.

... This time small finds abound and include a variety of non-local (foreign) ceramics, a large number of semi-precious stones and glass beads (over 3,000), copper coins, most of them in a corroded condition, iron, copper, gold and tin artefacts, cameo blanks, spindle whorls, terracotta lamps and so on.

Pattanam, located 5.5 miles (9 km) south of Kodungallur, is said to have been first occupied around 1000 BC and continued till the 10th century AD.

... The evidence points to the possibility that the site had the benefit of the services of a large number of artisans and technicians but not necessarily residents on the site. The plethora of artefacts and structures indicate that this site could not have been provisioned without a skilled workforce.

... The workforce comprised blacksmiths (large quantity of iron objects such as nails, tools etc.), coppersmiths (copper objects), goldsmiths (gold ornaments), potters (huge quantity of domestic vessels, lamps, oven and other terracotta objects), brick makers, roofers (large quantity of bricks and triple grooved roof tiles), stone bead makers, lapidaries (as indicated by a variety of semi-precious stone beads, cameo blanks and stone debitage) and weavers (signified by spindle whorls).

So here we have a site dating back to 1000 BC – well before the heydays of classical Greece and Rome – which contains central Mediterranean artefacts. Moreover some of these items, such as the gold ornaments, terracotta objects and lamps and stone beads, appear to be remarkably similar to those found in the Uluburun wreck; this will be considered in more detail later.

As the Vice-Chancellor of Tamil University, Mr M. Rajendran, said in a KCHR press release:

I am personally surprised to see the huge quantity of glass and semi-precious stone beads at Pattanam which goes well with those at the Kodumanal site, excavated by the Tamil University. The evidence unearthed at Pattanam definitely points to connections of the region with the Mediterranean world, South-East Asia and Sri Lanka.

I should add at this point that teak provisionally dated to the 2nd millennium BC and originating in Kerala has been found in the foundations of the Mesopotamian city of Ur.

VISIT TO COCHIN (KOCHI)

We selected Cochin as our base for a research expedition as it is on the backwaters of central Kerala, at the mouth of an estuary of the Periyar River. It is also near Pattanam, which is thought to be the site of ancient Muziris. From here we could explore both the coast and the interior by travelling up the Periyar.

Never will I forget arriving in Cochin at the Old Harbour Hotel at dusk on a warm tropical day. The hotel was once the Portuguese viceroy's palace. It has been superbly converted; the bedrooms have the original teak floors, with planks over 10 metres (33 feet) long and 1 metre (3 feet) wide.

Our bedroom overlooks the old harbour, which is framed by Chinese fishing nets perched along the length of the coast. These resemble great spiders whose front legs lean forward into the sand when the net (beneath the spider's belly) is lowered into the sea. There it rests for some five minutes before a counter-weight is lowered between the spider's rear legs. This raises the front legs and the net – now full of fish. We rush down and buy a big pomfret and a fresh snapper, which the hotel cook grills for our supper.

We eat in the central courtyard, which is centred around a large pool on which float purple water lilies. The air is redolent with frangipani. A sitar and an Indian flute play in the background. The courtyard is shaded by a huge rain tree from which enormous bats flit across the sky. A mango tree sprouts pink orchids; bamboos sway in the sea breeze; jasmine, jack fruit, spider lilies and heliconia surround us as we drink our chota pegs. Later the honk of ships passing downriver to distant lands lulls us to sleep – life could not come better than this; we are so incredibly lucky to experience such a day.

On our second night we ask to move temporarily next door to experience life in the Koder House. It was built by the Koders, prominent Jews, whose forebears traded in Kerala long ago. Our bedroom is 20 metres (65 feet) long, with the same enormous teak floorboards as in the Old Harbour Hotel. The family patriarch,

Samuel Koder, built the house on top of a former Portuguese palace. His house hosted various presidents, prime ministers, viceroys and ambassadors. Sam Koder's 'open house' every Friday was a focal point of the Raj establishment weekly social round. Kay Hyde, an 81-year-old from that era, describes Friday Open House:

> I met so many bigwigs there. Because of the Koders I met Benjamin Britten the famous composer; Peter Pears the singer; Princess Margaret of Hesse, sister of the Duke of Edinburgh; and Maharani Gayatri Devi. In those days Jew town at Cochin was full of Jews – almost all emigrated with the creation of Israel of 1948.

Lord Curzon, the British Viceroy of India, wrote an open letter of greeting to the Jewish community:

> Cochin and its people owe much to you. The memory of your early association with this country has always been pleasant. It is recorded by historians that your people began to visit this coast as early as the days of King Solomon [10th century BC] and they formed one of the earliest links binding East and West, fast with each other.

What I had picked up was an intriguing detail, in the form of the Jewish name for their home town. The Jews called their old settlement 'Shingly', an echo of the ancient name for Muziris. The fame of this settlement, ruled by a Jewish king, spread far and wide. To quote Rabbi Nissim, a 14th-century poet:

> I travelled from Spain,
> I had heard of The City of Shingly
> I longed to see an Israel King
> Him, I saw with my own eyes

Shingly became a haven for the Jews; their attachment was so strong that until relatively recently the Jewish custom worldwide was to put a handful of Shingly sand into any coffin, together with a handful of earth from the Holy Land.

THE BACKWATERS

By now Marcella and I have a reasonable feel for Cochin. Before our journey into the interior to find the fabled spices of the foothills of the Western Ghats, we decide to investigate the backwaters around the estuary of the Periyar River. This is the area where Muziris and other prehistoric ports emerged. We take a punt arranged by the Cochin Tourist Board. Our first impression is one of monotony – in whichever direction one looks there are coconut palms of the same shape and height for mile after mile. A navigator approaching the coast on the monsoon winds would find identifying the port area a very difficult task – he would have to know the precise latitude of the place or risk missing it altogether.

The excited foreign mariner arriving off the Periyar estuary would find a string of sandbanks sheltering lagoons from the sea on which to anchor, fresh water from the streams emptying into the back-waters, fish and fruit of every description as well as wild ducks, pheasants and quail.

JOURNEY INTO THE INTERIOR

We shall travel to the foothills of the Western Ghats up the Periyar River. These days it has channels which enter the Indian Ocean at Cochin and Pattanam (Muziris). Over the centuries the river has changed direction at its estuary, on account of the silt continuously brought downriver by the monsoon rains. Today the big estuary is at Cochin, 2,000 years ago it was at Pattanam. As the river changes course so do the ports where sea meets river. Up until Alwaye the river is as wide as the Thames at London or the Hudson at New York. Pattanam/Muziris is at the northern end of Vypin Island, which can easily be reached from Cochin by the ferry opposite the Old Harbour Hotel – cost three rupees (approximately four cents). Above Alwaye the river runs in a remarkably straight course, fringed by palm trees – just like the river in the frescoes of Thera (described in chapter 2).

For the first two hours we travel beside the river. There are still wild elephants in the forests here as well as tigers and leopards in the Periyar Nature Reserve. The land is rich in fruit – mangoes, bananas, papayas, pomegranates, tree tomatoes, passion fruit and chikoos. The river, according to Roman accounts, is rich in pearls. The woods were famous for jungle fowl, francolins (a type of

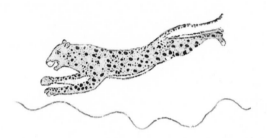

partridge), peafowl and wild deer – once again plenty of food for seafarers travelling upriver.

We have come at the tail end of the monsoon in mid-October to the festival of lights – Diwali. By the time we reach the foothills of the Western Ghats it is raining, or rather there is a thick wet mist dripping off the trees and shrubs. Josey, our splendid driver, suddenly stops and we get out. He shows us a tree surrounded by a climbing vine, with bunches of small green berries – pepper! Within a 5-square-yard (4 square metres) stretch of forest are wild coffee, cocoa, pepper and cardamom bushes or trees. It is the first time I had seen cardamom growing in the wild – cardamom grows on little stalks at the base of the shrub whose shape resembles a bamboo but with thicker leaves. Josey tells us that these foothills of the Western Ghats have perfect conditions for pepper and cardamom: just the right amount of shade and moisture, the right altitude – 600 to 1,500 metres (2,000 to 5,000 feet) – ideal soil conditions and a suitable all-year-round temperature.

The *Rough Guide to Kerala* has a very good summary that illustrates the lure Kerala would have had for any enterprising foreign trader:

Aromatic spices have been used to flavour food, as medicines, and in religious rituals for many thousands of years in Kerala. Traders from Sumeria first sailed across the Arabian Sea in the 3rd millennium BC in search of cinnamon and cardamom – centuries before the Romans mastered the monsoon winds and used them to reach the Malabar pepper, the 'black gold' prized in Europe as a taste enhancer and preservative ... few aspects of life in Kerala have not been shaped by spices in some way. Eaten by every Kerallen every day, they're still a source of export dollars and a defining feature of the interior hills, where they are grown in sprawling plantations.

The mariner approaching Kerala 4,000 years ago would have arrived at a coast rich in fish, fruit, game, water and building materials. Sailing up the Periyar River would have brought him into contact with elephants, falcons, leopards and pearls. Travelling further on foot he had the world's richest spices at his feet – pepper and cardamom, literally worth their weight in gold. All he had to do was to pick them or buy them dirt cheap. They were easy to ship and could be sold in the spice bazaars of Cairo or Mesopotamia for fifty times what he had paid in Kerala. Prodigious wealth was there for the taking. No wonder mariners have crossed the Indian Ocean for millennia with the trade winds to reach fabled Muziris and her black gold.

Arriving at the clubhouse of the High Range Club, a hill station above Munnar Town, is like entering a 1920s time warp. The walls of the formerly 'Men Only' bar are covered with photographs of man-eating tigers shot by members and of members who have played 'a hole in one' on the golf course. Marcella and I unwind over a chota peg while playing billiards. The golf course is flooded, so we opt for squash. We are the only visitors, spoilt by being surrounded by old-fashioned retainers in the livery of clubs of long ago. Tiffin has to be ordered an hour in advance and we are given strict instructions about dress – ties and dinner jackets on Saturday night. Drivers are not allowed to park on the gravel. All *ayas* have to leave by six p.m. and are not allowed in guest bedrooms before ten a.m. The setting is superb; we are surrounded by mountains covered in tea bushes, all clipped so even that the slopes resemble a vast sloping billiard table. By nightfall the temperature drops to near freezing and we fall asleep to the sound of distant macaques.

MINOANS IN KERALA?

The Uluburun wreck contained a number of items which could have come from India, including cowrie shells that the 'Beyond Babylon' exhibition had identified as from the Indian ocean. The elephant tusks in the Uluburun wreck could also be from Africa rather than India, but some animals shown in the frescoes at the Minoan home base of Thera are of certain Indian origin.

Admittedly, the leopards seen on the Thera frescoes could have been from Africa as well as India. The straight, palm-fringed river on the miniature fresco could be the Nile just as well as the Periyar. However prehistoric tusks found beneath Middle Minoan II tombs in Crete have been positively identified as being from Indian elephants.

Certain botanical specimens found in Mesopotamia and the Mediterranean in the 2nd millennium BC are uniquely Indian and can only have been carried from India to the Middle East by ship. Foremost among these is teak. As we have seen in our journey up the Periyar River into the interior, Kerala's cool rainforest on the foothills of the Ghats provides ideal conditions for both teak and sandalwood trees. The trade in teak between Kerala and the Middle East is proven by finds at both the prehistoric Egyptian port of Berenice and in excavations at the Mesopotamian city of Ur.

Archaeologists from UCLA (University of California, Los Angeles) and the University of Delaware excavating at Berenice have found extensive evidence of sea trade between the Far East, India and Egypt. They reported:

Among the buried ruins of buildings that date to Roman rule, the team discovered vast quantities of teak, a wood indigenous to India

155

and today's Myanmar but not capable of growing in Egypt, Africa or Europe.

... The largest amount of wood we found at Berenice was teak ...

The archaeologists uncovered the largest amount of Indian goods ever found along the Red Sea, including the largest cache of black pepper from antiquity: 7 kilos (16 pounds).

Peppercorns of the same vintage have been excavated as far away as Germany. The team also found Indian coconuts and batik cloth, sapphires and glass beads which appear to have come from Sri Lanka and beads which appear to have originated in Java, Vietnam or Thailand. Even more curious, the remains of cereals and animals indigenous to sub-Saharan Africa were found, pointing to a three-way Indian Ocean trade – southern Africa to India, India to Egypt, Egypt to southern Africa. Roman texts that address the costs of different shipping methods describe land transport as being at least twenty times more expensive than sea trade.

This international trade evidenced at Berenice is mirrored by that of Muziris/Pattanam, where the earliest dating (charcoal fragments trench II) is 1693 BC to 509 BC. This trade between the Mediterranean and the Indian port of Muziris in described in Tamil Sangam poems:

... Where the splendid ships of the Yavanas bring gold and return with pepper beating the foam on the Periyar ...

Ptolemy in *The Periplus of the Erythrean Sea*, and the Greek historian Strabo, have written numerous descriptions of trade between the Malabar coast and the Western world.

Professor Cherian and colleagues state:[29]

The excavation findings suggest that Pattanam had a key role in the early historic Indian Ocean trade. The archaeological evidence vouches for its cultural linkages with the Mediterranean, Red Sea, west Asia, Ganga Delta, Coromandel Coast and South East Asian regions ...

... An interesting possibility emerging out of the present study could be the possibility of tracing back the antiquity of external contacts to the pre-Roman period. The presence of non-European especially Nabatean (?) and West Asian variety of pottery could be another indication for maritime activities at Pattanam in the pre-Roman era.

Professor Cherian's work is corroborated by Emeritus Professor John Sorenson and Emeritus Professor Carl Johannessen's painstaking putting-together of the evidence for extensive maritime activity between India and Egypt, India and America (trading cotton) and America and India (trading corn) dating back four millennia. Their research shows that sea trade – pioneered by the Minoans and later adopted by the Phoenicians and Romans – between America and India was commonplace a full 3,000 years before Columbus.

By now I was convinced of the Minoan presence in Kerala during the Middle Bronze Age. I looked about for their 'signatures' – beyond those of the bronze weapons I already knew were found buried in the area of the Junma catchment of the Ganges. Could I find any 'signature' Kamares pottery, distinctive or amber jewellery, frescoes, or particularly unusual customs, such as the practice of bull-leaping?

The pottery excavated so far comes in shards and to my unpractised eye does not conclusively 'shout' Kamares ware. There was some interesting evidence of bull-leaping in the annual celebrations of '*Jellikata*', a ceremony during which youths lead the bull to '*encearos*' then attempt to somersault on to its back. I was struck by the Spanish-sounding terminology, but I am no linguist and certainly not when it comes to Indian dialects. This strange custom is still practised and young people get gored, just as they do in Spain today (as described later). However, though bull-leaping appears an odd and unlikely custom, this in itself could be a coincidence. It seems less so when you consider that this is a Hindu society, where cattle are considered holy, objects of veneration. This custom has clear similarities to those ancient Crete.

I felt confident that Minoans used to travel upstream on the

Periyar River for pepper, cardamom, teak and perhaps occasionally exotica such as leopards and monkeys. It was hard, though, to keep tabs on information from India, so once we'd returned to England I retained an agency to search English-speaking Hindi and Malayalam newspapers for relevant leads.

We did not have long to wait although the lead, when it came, was from a totally unexpected direction. On 10 June 2009 *The Hindu*, India's national newspaper, ran this headline in its online edition: 'Prehistoric Cemetery discovered in Kerala.'

Triruvananthapuram: Archaeologists have discovered a pre-historic necropolis (cemetery) with megalithic cairn circles dating back 2,500 years ... a woodhenge-like ritual monument and a site of primitive astronomical intelligence at Anakkara, near Kuttippuram in Malappuram District [about 93 miles (150 kilometres) north of the Periyar River estuary].

According to the press report, interred objects suggested that the find was around 2,500 years old.

We could also trace some broken pieces of an unidentified copper object. These artefacts could be indicative of the earliest trade contacts of the region ...

It continued:

... Similar posthole finds at certain necropolis sites of Anatolia, Syria, Greece, London and so on have been reported in connection with the Neolithic as well as Bronze Age cultures of secondary burial practices. Nevertheless, our traces of erecting posts in their holes using insights of experimental archaeology, have turned out to be quite interesting and revealing. The posthole alignment looks exactly like that at Woodhenge, in England. The holes of uneven sizes, big and small interspersed, in a strikingly wide open site ideal for star watching probably indicate patterns of heavenly bodies and are suggestive of primitive astronomical intelligence.

This was totally unnerving. The dates seemed to match, closely following the time that the Minoans would have been at their most active. How strange that a European-style prehistoric ceremonial circle should have been found in Kerala and that it should be so close to the river that I was sure the Minoans had traversed.

The newspaper's assertion that the Keralan wooden circle was for star-watching is what held my particular attention. Wherever they were, the Minoans would have needed to navigate their way back. And, like Homer's hero Odysseus, they would have had to do that via the stars. Perhaps they had relied on observatories. Perhaps the Minoans had even built those observatories.

18

THE TRUTH IS IN THE TRADE ...

I have long been convinced that world travel has had a much more complex history than historians allow. And as I was researching the products and produce that the Minoans brought through Egypt and into India – and back again – I was uncovering so much unexpected information I hardly knew where to begin to unravel all the leads.

In India, I'd discovered that beautiful rock art carvings and paintings of American bison have been found on the borders of Kerala and Tamil Nadu, near the point where the Periyar River rises. They were dated to the 2nd millennium BC. I had to ask myself how Keralan artists of 4,000 years ago had any knowledge of American bison.

The answer seemed to be that the connecting tissue between all these strong, growth- and wealth-hungry cultures was the enterprising and dauntless Minoans. To take this any further, I needed to find whatever proof I could that there had been transatlantic contact millennia ago. Back at my desk, I struck gold when I turned once again to Emeritus Professors John Sorenson and Carl Johannessen. Both academics have documented a considerable amount of intercontinental trade before Columbus, as has Dr Gunnar Thompson. Researching their carefully documented material over painstaking decades, they have produced a huge amount of detailed information on the 2nd millennium BC, which has been contested by those who seem unable to ditch the paralysing historical convention that only Christopher Columbus could have discovered America.

From their work I found that there were largely six categories of goods for which there is substantial proof of trade. That exchange was between India, Egypt, the Minoan lands and North America, in the era of Hatshepsut's expeditions to the East: that is, during the 3rd and 2nd millennia BC. They have documented: (1) maize transported from the Americas to Egypt and India; (2) descriptions of the origins of this maize given by Indian and African peoples; (3) cotton taken from India to the Americas; (4) tobacco and drugs transported from the Americas to Egypt via Thera; (5) squashes and fruit shipped from America to India; (6) gourds exported from India to America. This will, of course, only be relevant if it can be demonstrated that it was Minoan ships in particular that traded with the Americas as well as with Egypt and India.

To concentrate on a few specifics that clearly illustrate local Indian knowledge of 'exotic' produce: in a highly detailed work of scholarship, Professors Carl L. Johannessen and Professor Anne Z. Parker contend that stone carvings of maize exist in at least three pre-Columbian Hoysala stone temples near Mysore (see second colour plate section). Professor Johannessen has also found the sunflower, another New World crop, in pre-Columbian Indian temple sculptures.[30]

Their conclusions are supported by a separate research project undertaken by the Indian botanist Professor Shakti M. Gupta,[31] of Delhi University, who agrees:

Different varieties of the corn cob (Zea Mays Linn) are extensively sculpted on the Hindu and Jain temples of Karnataka. Various deities are shown as carrying a corn cob in their hands, as on the Chenna Kesava Temple, Belur.

Professor Gupta continues:

The straight rows of the corn grains can be easily identified. In the Lakshmi Narasimha temple, Nuggehalli, the eight-armed dancing Vishnu in his female form of Mohini is holding a corn cob in his left hand and the other hand holds the usual emblems of Vishnu . . . .a 12th century sculpture of Ambika Kushmandini sitting on a lotus seat under a canopy of mangoes holds in his left hand a corn cob . . . .

The number of instances is overwhelming. Professor Gupta identifies sunflowers, pineapples, cashews, custard apples and monstera – all of them New World species – in the pre-Columbian art of India. Carvings depicting the sunflower, a native of Central and South America, are found for example in the Rani Gumpha cave, Udayagiri (2nd century BC). As Professor Gupta reports, a pineapple is clearly depicted (see second colour plate section) in the Udayagiri core temple, Madhya Pradesh (5th century AD); a cashew in the Bharhut stupa (2nd century BC) and monstera – a climber native to Central America – in Hindu and Jain temples in Gujarat and Rajasthan (11–13th centuries BC).

As for dating the era when maize first appeared, Professor Gupta implies that it may have been in existence long before it appeared in the carvings described above:

> ... it is quite conceivable that maize was present in the subcontinent [India] for many centuries before the Hoysala dynasty [described by Professor Johannessen], and that distinctively Asian varieties were developed early on.

Indian and African people believed maize was first brought to their countries 'from Mecca' in the case of Indians and 'from Egypt' in the case of Africans. These descriptions would make sense if my suspicion that the Minoans had known exactly how to get to America was true. Perhaps now was the moment to investigate the little *Lasioderma serricorne* beetle in more detail.

THE TOBACCO BEETLE

If, as it appears, the Minoans carried tobacco from the Americas to Egypt, then there should be evidence of American tobacco in Crete itself or on Thera, the principal Minoan base. Finding such evidence may be complicated by the destruction visited on both islands by the volcanic eruption.

There is, nevertheless, such evidence – in the form of the tobacco beetle, which I suspected would also be found in ancient Egypt. The first specimen I had encountered was buried beneath the 1450 BC volcanic ash of a merchant's house in the Minoan town of Thera, modern-day Akrotiri.

Lasioderma serricorne was indigenous to the Americas. Yet, as Sir Walter Raleigh, who brought tobacco to 16th-century England as a prize for his queen and sponsor Elizabeth I, would have testified, the plant is not indigenous to Europe. Nor did it grow there in 1450 BC.

The life-cycle of this little beetle, averaging forty to ninety days, is highly dependent on temperature and food source. Females lay ten to a hundred eggs in the tobacco and the larvae emerge in six to ten days. The larvae cannot hatch below seventeen degrees Celsius and they die when the temperature falls to four degrees Celsius. In short, the beetles can only breed in warm conditions. In the right season, Minoan ships would have had a warm hold, with sailors sleeping in it off watch.

After a period of further desk research, I knew that this tobacco beetle had also been found in the tombs of pharaohs who had clearly smoked the weed – Anastase Alfieri reported finding them in King Tutankhamun's tomb (1931 and 1976) and J.R. Steffan (1982) in the visceral cavity of Ramses II. Alfieri (1976, 1982) reported specimens from Alexandria, Cairo, the Nile Delta and the Fayum and Luxor regions – all places visited by the Minoan fleet. So I believe that the carriage of tobacco and the tobacco beetle is another 'signature card' of Minoan voyages to the Americas and back; as well as to Egypt and the Mediterranean. It is a card just as distinctive as the Minoans' cult of the bull. It also dates at least some of their travels to the specific era between Tutankhamun and Ramses II. The date range would have been 1336 BC (Tutankhamun) to 1260 BC (Ramses II at the age of about 38), and in Thera to before 1450 BC (the possible date of the volcanic eruption) – a 200-year spread.

Yet there was another obstacle in the way of this theory. It came very literally, in the shape of Africa. From Crete's position in the Mediterranean; from the Egypt of the pharaohs; from India, it was all the same. The vast looming land mass of Africa is dominant on any modern map. From both Egypt and India, the so-called 'Dark Continent' is right in the way of anyone attempting to get to the Americas. It appeared to me that there was only one way that North American produce could ever have reached India. The Minoans can only have gone west, not east. They must have braved the perils of the North Atlantic ocean and breached the Straits of Gibraltar, passing the landmark that the ancient Greeks, believing that the two magnificent peaks were the gates to the end of the world, called 'The Pillars of Hercules'. The pillars beyond which, according to Plato, lay the lost realm of Atlantis.

Notes to Book II

1. Homer, *Iliad* 16.221–30; 23.196, 219, trans. Robert Eagles. Penguin Classics, 1998

2. G. F. Bass, 'Cargo From the Age of Bronze', in *Beneath the Seven Seas*, Thames & Hudson, 2005

3. Cathy Gere, *The Tomb of Agamemnon*, Profile Books, 2006

4. Tim Severin, *The Jason Voyage*, Hutchinson, 1985

5. Ibid. p. 161

6. Robert Graves, *The Greek Myths*

7. D. Grimaldi, 'Pushing Back Amber Production' in *Science*, 2009 p 51–52

8. Hans Peter Duerr, *GEO Magazin*, no. 12/05. Dienekes 8/2008

9. Joan Aruz, *Beyond Babylon: Art, Trade and Diplomacy in the Second Millennium BC*, Barnes & Noble, 2008

10. A. Hauptmann, R. Maddin, M. Prange, 'On the Structure and Composition of Copper and Tin Ingots Excavated from the Shipwreck of Uluburun' in *Bulletin of the American Schools of Oriental Research*, no. 328, pp. 1–30

11. William L. Moran, *The Amarna Letters*, John Hopkins University Press, 1992

12. Joan Aruz, *Beyond Babylon*, p. 167

13. Bernard Knapp, 'Thalassocracies in Bronze Age Eastern Mediterranean Trade: Making and Breaking a Myth', in *World Archaeology*, vol. 24, No.3, Ancient Trade: New Perspectives, 1993

14. Rodney Castleden, *Minoans*, p. 32

15. D. Panagiotopoulos, 'Keftiu in context: Theban tomb-paintings as a historical source', *Oxford Journal of Archaeology* 20, 2001, pp. 263–4

16. Leonard Woolley, *A Forgotten Kingdom*, Penguin, 1953, pp. 74–75

17. The Thera Foundation 2006, Heaton 1910; 1911; Forbes 1955, pp. 241–242; Cameron, Jones and Philippakes 1977; Hood 1978, p. 83; Immerwahr 1990a, pp. 14–15. *Aegean Frescoes in Syria-Palestine*

18. Joan Aruz, *Beyond Babylon: Art, Trade and Diplomacy in the Second Millennium BC,* Barnes & Noble, 2008

19. From USA Today, 5th March, © 2006 USA Today. All rights reserved. Used by permission and protected by the copyright laws of the United States. The printing, copying, redistribution or retransmission of this content without express written permission is prohibited

20. André Parrot, 1958a, 165 n.2, *Samaria: the Capital of the Kingdom of Israel,* SCM Press, 1958

21. The Teaching Company User Community Forum Index, Alexis Q. Castor. 'Between the Rivers: The History of Ancient Mesopotamia,' Teaching Co. Virginia, USA

22. Jack M. Sasson, in *Beyond Babylon: Art, Trade and Diplomacy in the Second Millenium B.C.* Yale University Press, 2008

23. Zephaniah 2:13 (Destruction of Syria and Nineveh)

24. Homer, *Odyssey*, trans. Samuel Butler, Longmans, 1900

25. Enuma Anu Enlil 17.2

26. The Canal Builders, Payn, Robert, MacMillan, New York, 1959

27. Herodotus, *Histories*, trans. George Rawlinson, Penguin Classics, 1858. p. 109

28. Paul Lunde, *The Navigator Ahmed Ibn Majid*, Saudi Aramco, 2004

29. Professor Cherian and colleagues, 'Chronology of Pattanam: a multi-cultural port site on the Malabar coast'

30. 'Maize Ears Sculptured in 12th and 13th Century A.D. India as Indicators of Pre-Columbian Diffusion', *Economic Botany*, 43

31. Shakti M. Gupta, *Plants in Indian Temple Art*, B.R. Publishing, 1996

BOOK III
||||||||||||||||||||||||||||||||

JOURNEYS WEST

19

NEC PLUS ULTRA:
ENTERING THE ATLANTIC

Helen of Troy's face is said to have launched 1,000 ships. Plato, meanwhile, claimed that the kings of his fabled Atlantis had 1,200 ships.

Even by the time of the Trojan War, Homer was still able to describe ancient Crete as the provider of eighty ships for the joint campaign to avenge Paris' abduction of fair Helen. We've established that the Minoans were a mighty sea power until the volcano struck. Thera alone, at the time of the admiral's fresco, seems to have had at least ten ships, and possibly more. Minoan power and reach expanded steadily as the Bronze Age developed.

I was now searching for the answers to two major questions. The tobacco beetle *Lasioderma serricorne* must have been brought to Thera somehow: exactly how? I also suspected that at least some of the near-pure copper found in the Uluburun wreck originated at Lake Superior. How could two so very different items from America have surfaced so far away, in the exotic island culture of the ancient Mediterranean? Geography dictated that it was almost impossible that anything from North America could have reached Minoan Crete via Egypt, or even through India. There was no other answer: sailors, not necessarily the Minoans themselves, must have crossed the Atlantic – and to meet up with them, the Minoans must have made their way west as well.

Yet to brave the unknown, voyaging through the Pillars of Hercules, which the ancient Greeks believed held up the very sky, would have taken real courage. There is much dispute about the location of the pillars – and indeed, about what exactly they were.

Although some scholars believe that the mythic hero Hercules performed his labours in the Greek Peloponnese, and dispute that the pillars lay between Spain and Morocco, there is an overwhelming amount of evidence in favour of the Straits of Gibraltar.

What would have tempted the Minoans to sail westwards in the first instance? Surely the answer has to be something to do with finding, or producing, the substance that ignited the whole age – bronze? The Uluburun wreck, much smaller than the ships shown in the Thera frescoes, carried 10 tons of copper in its hold: a vast amount.

Crete, the birthplace of Minoan civilisation and the lynchpin of the Bronze Age, had no copper in usable amounts. The island had some small amounts of surface copper, but not enough to fuel the huge amount being smelted and worked on a grand scale in Minoan Crete, not least in the palaces themselves. Intriguingly, at Chrysokamino on the northeast coast of Crete there is evidence of smelting activity stretching as far back as c.4500 BC. Crete had no tin. It had to be imported by sea.[1] The surrounding, colonised islands were also engaged in smelting: after all, the magical metal of bronze was the crude oil of its day, as valuable as the gold that made Croesus rich. So where did the raw materials that fuelled the Minoan brilliance in bronze actually come from?

Copper, and later on in time the finished bronze, must have been imported by sea.[2] The nearest copper mines were at Lavrion in mainland Greece and on the island of Kythnos in the Cyclades.

||||||||||||||||||||||

Bronze, to recap, ideally contains around 90 per cent copper. The other 10 per cent is made up of tin or arsenic. Mediterranean traders could definitely find arsenic-alloyed copper, but no tin. Almost all copper ores contain some level of arsenic. Arsenic ores are more common than tin ores and are generally found in mines in western Asia. And, in fact, since 3000 BC, Cretan and western Mediterranean bronzes had been made with arsenical copper. So there was no real need for tin – or so it might appear on the face of things. The deadly drawback in working with arsenical bronze, however, is a major one: the risk of death by poisoning. Perhaps it was partly as a result of this danger that bronze-making carried with it a strange allure.

All of the evidence suggests that in the Bronze Age the actual bronze-maker, or master, was respected – lionised, almost – as if he were a magician. In some cultures, the actual bronze-making process was top secret: given the alloy's huge value, the knowledge may also have been regarded as sacred, and could even have been restricted in some societies to kings or shamans. Yet king or not, the unfortunate Bronze Age coppersmith's lot would have been an unhealthy one. For all his status, he could not have avoided breathing in the fumes of arsenic as he heated, cast and hammered the hot arsenical bronze.

The smiths' illnesses have come down to us through legend – both the Greek god Hephaestus and his Roman counterpart Vulcan were crippled. So this must be at least one reason why tin bronze had become the Minoans' metal of choice by about 2600 BC. When it comes to making desirable objects such as swords, tin bronze also has the technical edge; while arsenical copper melts at 1,084 degrees Celsius, tin bronze melts at 950 degrees Celsius, making it much less prone to cracking in the mould.

There were some sources of base metals in established trade areas near to Crete. Göltepe in the Taurus mountains, in what is now Turkey, had tin mines producing substantial quantities of ore from 3290 BC until 1840 BC, when the town was sacked and tin production ceased.[3] So from 2600 BC up until 1840 BC there was theoretically a tin source readily available in the Mediterranean. Yet the records of King Sargon of Akkad say that Minoan ships had been sailing to the Atlantic to obtain Iberian and English tin since 2350 BC. Once again, why bother to look westwards in the five centuries *before* Göltepe ceased tin production?

The answer, I think, lies in the third material required to make bronze – wood. To produce just 1 kilogram (2 pounds) takes around 300 kilograms (660 pounds) of charcoal, smelting 30 kilograms (66 pounds) of ore. The Bronze Age may have opened up almost limitless opportunities for civilisations to grow and develop, but like the Minotaur it was a ravenous beast to feed, devouring acres and acres of trees. The evidence is in the *Epic of Gilgamesh*.

Archaeology owes a lot to a Victorian Londoner with the unassuming name of George Smith. Only a few months before, in my journey to the Near East, I'd sought to the great city of Nineveh to track down Aššurbanipal's library. Where I had found only mounds of earth and ruptured walls, however, George Smith had found real words, real stories and real history – and in translating them, told us much of what we now know about ancient Mesopotamia and the Middle East, and the region's lost legions of precious trees.

Nineveh was first rediscovered by the French Consul General at Mosul, Paul-Émile Botta, in the 1840s. By the time the British adventurer Sir Henry Austen Layard had later uncovered the fabled library, Nineveh was still less of a city than an ancient conundrum, shrouded in mystery. With archaeology still in its infancy, academics were at a loss. They could not interpret what had been found – although they knew they had at least one source of information: the Bible. Nineveh is first mentioned in the Old Testament, in Genesis 10.11. By the time we get to the Book of Jonah, though, things have gone very badly wrong for its citizens; God has brought down destruction upon a corrupt society. Rigorous historical knowledge about the city, however, was almost completely lacking. Smith is the rather unexpected figure who let us in through the Assyrian city's gates by translating the cryptic cuneiform inscriptions – and the stories – left us by the ancients. The translator was from a poor family, so he had received little education: and he certainly had no background as a linguist. Even so, he was both determined and dedicated. In 1876, Smith became ill and died while he was searching for more missing tablets around Nineveh. But not before translating the first written poem in history – the stone tablets that together form the *Epic of Gilgamesh*.

The epic tells us that in antiquity vast forests grew in the Middle East. That can be hard to believe today, when what we so often see, particularly on the news or in documentaries about the Middle East, is an unrelentingly barren and unforgiving landscape of hard earth, baked dry by the sun. Yet in the early part of the 3rd millennium BC, the mountain slopes of the region were covered with massive cedar forests. In the resource-hungry millennium before Christ's birth those millions of trees, including many of the much prized cedars of Lebanon, essential for building ships, were to disappear totally.

The real-life Gilgamesh was a Sumerian king of Uruk, living around 2700–2500 BC, who after a series of victories in war ruled the whole of southern Mesopotamia. The middle stories of the Gilgamesh epic warn about the perils of deforestation.[4] Approximately 4,700 years ago in Uruk, a city-kingdom in southern Mesopotamia, the king and his companion set off to Lebanon to cut down her famous cedars. They incur the wrath of the forest god Humbaba, who in his turn punishes Gilgamesh. The story illustrates the fact that Mesopotamia did not have enough wood to support its flourishing Bronze Age civilisation. As Mesopotamia constructed thousands of roads, cities, canals and palaces using bronze, the forests were felled to fuel the furnaces. These new cities needed public buildings, palaces and cisterns in which to store water in the dry summer months. For these they also needed cement, plaster, brick and terracotta – all of which required fuel for their manufacture.

It's intriguing that Plato also talks about an environmental crisis. In his dialogues he describes the consequences of deforestation on the Acropolis:

> The land was the best in the world ... in those days the country was fair as now, and yielded far more abundant produce ... there remained only the bones of the wasted body, as they may be called, as in the case of the small islands. All the richer and softer parts of the soil having fallen away, and the mere skeleton of the land being left ... there was [long ago] abundance of wood ... the traces still remain.

The problem of deforestation became so serious that around 1750 BC King Hammurabi of Mesopotamia set up laws against unauthorised tree-felling.[5] So I was beginning to suspect that war may not have been the only reason why tin mining did not start again in Göltepe after it had been sacked – perhaps it had used up its wood supply.

The difficulties involved in replacing and replanting timber are much more acute in dry Mediterranean countries than in wet northern Europe, where trees grow relatively quickly. In the Mediterranean, rain is restricted to winter and tends to be sudden and violent. The irony is that without trees to anchor it, the soil itself gets depleted – just, in fact, as Plato had described.

> ... winter rains on steep deforested slopes quickly degrade the soil by washing it downhill. Seedlings have difficulty in re-establishing the forest, especially after clean cutting, and the soil quickly degrades to the point that pine forest cannot recover ... The tremendous tonnage of ancient copper slag on Cyprus suggests that the Cypriot copper industry collapsed around 300 AD, simply because the island ran out of cheap fuel. The slagheaps suggest a total production of perhaps 200,000 tonnes of copper – and that in turn suggests that fuel equivalent to 200 million pine trees were cut to supply the copper industry, forests 16 times the total area of the island.[6]

The problem would have been exacerbated by the conditions in the dry eastern Mediterranean, because the coastal cities and nearby Mesopotamia and the Levant were all developing rapidly and doing so in parallel with each other.

If Mari's glorious 5-acre palace is anything to go by, then the Bronze Age was nothing if not architecturally ambitious. Meanwhile Egypt needed incalculable amounts of bronze to make the tools for its astonishing, relentless programme of pyramid and temple construction. Not that much later in time, the Mycenaeans would construct entire roads leading into magnificent edifices that blazed with gleaming copper and bronze.

Later on in history, the problem becomes still more acute. The Athenian fleet that defeated the Persians at Salamis was built from timber that came from the Balkans and southern Italy, rather than Greece, because Athens had so depleted her own forests. T.A Wertime estimates that the Laurion silver mines near Athens consumed an awe-inspiring 1 million tons of charcoal and 2.5 million acres (1 million hectares) of forest.[7] The mines perhaps did not stop production because of a shortage of ore, but because imported fuel costs had risen so much that they were no longer economic.

Seen in that light, developing trade with the western Mediterranean and possibly even further – to the wet, Atlantic shores of Europe – made absolute sense. Not only were there abundant supplies of copper and tin in Iberia – but timber was plentiful. There was a lot of evidence on the museum shelves at Thera that suggested, at the least, mutually profitable trade. The Minoans would export their finished goods in return for the natural resources of wider Europe: much like Europe's relationship with Africa today. Bronze pins, for example, of the same type are found across Iberia, France and Britain. I'd seen the exact same design in the museum at Akrotiri. Were they exported from Thera?

In this scenario, Minoan shipbuilding expertise now becomes even more valuable. They did not have the metal ores on Crete. Paradoxically, this negative situation eventually turned out to be a strength for the Minoans; it meant that, latterly, large-scale smelting did not have to take place on Crete. If that had been the case they would still have had the wood to build their ships – and with that ability to travel came the opportunity to smelt plentiful ores *in situ*, where both the ore and the wood were readily available. They could have controlled the entire supply of bronze to the eastern Mediterranean. The Holy Grail was theirs.

At first, it seemed almost fanciful to think that the Minoans would roam that far across the globe. But bronze was the most powerful metal yet known to man. I knew there had been plentiful supplies of Cornish tin and Welsh copper in the British Isles. Now I discovered that experts believe that during the Bronze Age about half of the 'wildwood' or native forest disappeared from Britain: suggesting the possibility of smelting on an industrial scale.[8]

To achieve their goal the Minoans would have had to pass right through the Straits of Gibraltar, into the unknown. I thought of that eerie phrase, which ancient accounts say was once written in stone on the Pillars of Hercules:

Nec plus ultra (venture thus far, but no further).

Many hundreds of years later than the Minoans, the Greeks and the Romans were so scared of the Straits that there were dire admonitions not to go there. There were terrifying stories of petrifying monsters of the deep in the unknown ocean beyond. Perhaps the Minoans, whose rich and artistic culture was passed on to the Greeks, had also had their own fears?

Yet, drawing on my own knowledge of seafaring, I was beginning to piece together a picture of the ancient Minoan mariners as the greatest sailors the world has ever known. We tend to think of 'progress' as if humankind is in perpetual forward motion; as if knowledge, technology and culture always advance relentlessly, hand-in-hand.

Studying the Bronze Age gave the lie to that proposition: under the microscope of history I could see great civilisations rise and fall like waves breaking over a ship's prow. I was learning that knowledge, even advanced knowledge, is as slippery as the deck of a clipper; and it can be lost a lot more easily than it can be gained.

||||||||||||||||||||

The breezes in a Mediterranean summer often blow in the evening, because of the difference in temperature between the land and the sea. At dusk, the desert cools surprisingly quickly and the wind blows offshore to the warmer sea. Thus, I calculated, hugging the Mediterranean coast from Crete to the Straits of Gibraltar – setting sail after dusk and sailing for a few hours each night – should allow a galley with a decent sail to reach the Straits of Gibraltar in fits and starts even without the use of oars (see map showing Mediterranean winds).

Based on the three months that Severin's *Argo* took to run the 1,500 miles (2,400 kilometres) from Spetses to Georgia, against wind and current, I thought the voyage from Thera to the Straits would perhaps have taken a third longer – four months.

There is solid evidence that the Minoans travelled westwards across the Mediterranean. A tale told by an intriguing name: a settlement called Minoa. According to Rodney Castleden:

> There was a port called Minoa on the south-west coast of Sicily, which may have been a Crete-controlled trading station. Quite what the Minoans wanted from the West is unknown ... the Minoans needed tin to make bronze, and the sources of their raw materials are unknown. The tin may have come from Etruria, Bohemia, Spain, or even Britain ... A gold-mounted disc of amber found at Knossos may have come from the Wessex culture of southern England.[9]

Over the half-century since Spyridon Marinatos' astonishing discoveries on Thera, the documented evidence of Minoan settlements has grown and grown and the movement seems to have been westwards. Some of those settlements may have had the suffix 'Minoa' added to their names. Those client islands, or colonies as they may have been, are identifiable by features shared with Bronze Age Crete: such as an irregular street plan in the towns; the distinctive Minoan style of architecture; Minoan burial customs; and the introduction of Minoan pottery shapes and styles – including wares that were not imported from Crete, but show the local inhabitants adopting Minoan designs for themselves. Sometimes, there is also evidence of the introduction of Minoan religious rituals, such as ritual cups or figurines.[10]

There seems to have been a rapid growth in the Minoan trading empire between 1700 and 1500 BC. The settlement at Kastri on Kythera, which began in early Minoan times, is an early and small move westwards; by contrast, Minoa's presence in Sicily shows the Minoans venturing far into the central Mediterranean, moving ever closer to the Balearic Islands and Spain.

So this left me with a question. Could the Minoans have set up further foreign bases, just as they had done at Avaris (Tell el Dab'a) in ancient Egypt? If I was correct, and copper of the purity found on the Uluburun wreck could only have come from America, ships setting out across the Atlantic would need a base for repairs, provision and preparation for the serious journey ahead. They would need storage facilities and possibly labour. Could there have been permanent forward bases established in southwest Spain or Portugal?

If the Minoans had got as far as the Straits, I reasoned, it would have been logical for them to explore the lands they had discovered there. I'd read various books that hinted at Minoan interest in Britain. Logically, they would have traded with Spain well before they even got to the British Isles. By chance, I also happened to know something about the wealth of minerals in what was once known as Iberia. Later in history, the Romans had been eager to conquer the peninsula for that very reason. Today's Spanish term for a stream, '*arroyo*', comes from the Latin '*arrugius*' – meaning a gold mine. Both Spain and its neighbour Portugal would have held great riches for eager traders in search of bronze. Could I too strike gold? I decided that Spain would have to be my next stop in the journey.

20

A FOLK MEMORY OF HOME?

Once the Minoans had succeeded in sailing the Mediterranean, it was surely only a matter of time before they discovered Iberia, one of the most heavily mineralised places on earth. Beyond the Straits are two majestic rivers, both of which would have led the Minoans straight to all of the glittering prizes of this land – not only copper, but gold and silver.

I will never forget entering the Guadalquivir in HMS *Diamond*, when I was officer of the watch. The Atlantic coasts of Spain and Portugal are always memorable to seafarers; not least because of the smell of hay as you pass Cape St Vincent *en route* to the magic East; on the way home, turning north on passing Sagres, you savour the scent of pines.

It was 1958. Like the Rio Tinto, the Guadalquivir debouches into the Atlantic Ocean at the very southern tip of Spain and would have been readily visible to Minoan mariners had they dared to broach the Straits. The river is one of the longest in Spain. It flows west for 408 miles (657 kilometres), emptying into the Atlantic at Sanlúcar de Barrameda on the Gulf of Cádiz. We had been ordered to pay a goodwill visit to Seville, some 70 miles (112 kilometres) upriver. The Guadalquivir is so shallow that a ship of the *Diamond*'s size could only pass over the bar at Sanlúcar de Barrameda at high spring tide. Which happened to be exactly the same time that every year a sheer wall of water, a 'bore', surges up the river. We had to time our entrance to the minute.

There was a huge judder underneath us. Then good old HMS *Diamond* had to mount the bore and travel on its crest which, since it races along at more than 20 knots, is a hair-raising experience. It was like skiing on stilts. What's more, a big ship travelling at this kind of speed causes a vast wave of its own. In vain we flashed our lights and blasted our fog horn to warn the farmers who were pootling their way to market along the riverbanks. Time after time the wave caught them – fortunately with no casualties, bar the odd half-drowned, furiously braying donkey.

So I had a special fondness for the Guadalquivir and paid attention when I noticed magazine reports which said that a Bronze Age settlement had recently been discovered on its estuary. In the process of creating a beautiful national park, the Doñana, the excavators had uncovered an ancient human settlement. They had also found the wrecked timber carcasses of what could well be Bronze Age ships, in the swamps the Spanish know as 'Las Marismas'. It would be a long time before all of the archaeologists' detailed studies were complete, but initial carbon dating estimates from the French excavating team suggested that the remains dated from 2000 BC.

I also knew the large modern port of Huelva, about 50 miles (80 kilometres) north, which is fed by the Rio Tinto. I'd visited it in the 1950s, but now I saw a completely different side to what I'd always thought of as just a 17th-century port town. The town's origins were definitely ancient and there were still a number of Roman remains, I read. Yet what was most intriguing here was that a specialist in human prehistory, Martín Almagro Basch, had investigated a hoard of bronze artefacts found hidden in the area. Other archaeologically significant hoards had also been found in the region, notably the Leiro hoard, which was discovered by a fisherman in 1976 in the estuary waters of the Ulla River, near Leiro in Galicia.

The River, or Rio, Tinto is named 'the red' for a good reason: as it trickles past the ancient fortified walls of Spanish medieval towns it runs a symbolic blood-red. Copper. Copper ore everywhere. The sheer wealth of the minerals in this landscape of other-worldly greens, yellows and reds gave the area a fabled status in times of yore: these were, according to legend, King Solomon's mines.

IIIIIIIIIIIIIIIIIIII

I began to research, again using the British Library – and this time, the Web.

If the Minoans did come here, then initially at least they could have simply sifted the incredibly rich alluvial sediments for metallic ore. Was there any evidence that they had done so? Much later than the Minoans, ancient Greek and Roman writers such as Strabo (63 BC–AD 23) wrote that northwest Iberia was by then well known as a rich source of tin, as did Ptolemy.

Starting a survey of Spain from the north, I looked for as much evidence as I could. Professor Beatriz Comendador Rey of the University of Vigo believes that northwest Iberia could also have a rich archaeological heritage of metallic finds, but that the wet climate of this part of Spain makes dating and scientific evaluation problematic. I was interested in this area, as I had already come across a lot of evidence to suggest that the Minoans had reached the Baltic. If they had, they would have had to follow the coastline, and from here at the chilly tip of Vigo move north through the markedly colder English Channel.

Professor Comendador Rey links similar finds by Almeida with Bronze Age awls found on the nearby islet of Guidor Aredso. She also refers to the artefacts recovered from the River Ulla, mostly swords and spearheads dating from the Middle/Late Bronze Age. Taking all of these finds together, it seems beyond argument that they came from shipwrecks of the Middle/Late Bronze Age.

Moving south, the vast Rio Tinto copper, silver and gold mines were first worked during the third millennium BC.[11] Mark A. Hunt Ortiz dates the mines' initial period of use to around 2900 BC.[12]

It is worth pausing at this point: continuing up the Guadiana River (see map) would bring us to Évora. Ten miles (16 kilometres) west of Évora is the Bronze Age stone circle of Almendres Cromlech. I looked at some pictures I found of the circle on a website. 'How strange,' I thought. 'That picture shows a carving on one of the menhirs.' An enthusiastic amateur photographer had somehow got a shot that was at exactly the right angle to show something carved into the rock. To my eye, it looked like an axe. Thinking back to the newly discovered stone circle by the Red Sea and also to the 'woodhenge' of Kerala, I made a mental note to return to Évora, both in body and in thought.

Sailing northwestwards along the coast for another 30 miles (48 kilometres), from the Rio Tinto estuary, would have brought the Minoans to the Rio Guadiana, and what we would now call Portugal. Its waters were most likely just as red as the Rio Tinto, for just 40 miles (65 kilometres) upstream are the copper mines of São Domingos (see map). To this day the water is still contaminated by copper, a threat to the health of both people and wildlife in the region. Archaeological digs at the mine have found prehistoric tools, showing that the ores here were exploited more than 4,000 years ago.

The Rio Guadiana is navigable upriver for over 100 miles (160 kilometres) into the interior after Évora: the river continues past Badajoz to Cuidad Real and the romantic, gaunt and beautiful land of La Mancha.

During the past thirty-five years, the work of Spanish archaeologists in La Mancha has revealed what is probably the highest density of Bronze Age settlements in Europe. There are many massive stone complexes of large, permanent, fortified Bronze Age settlements. Their extent only became apparent after excavations had been carried out by the University of Granada in 1973. Survey work in Albacete has documented no fewer than 43 Bronze Age settlements and another 300 Bronze Age occupation sites. According to Dr Concepcion Martin:

> The concentration of surviving early 2nd millennium Bronze Age settlements in La Mancha has few parallels elsewhere in western Europe.

> Almost all our knowledge of the Bronze Age of La Mancha comes from recent excavations ... It is clear that La Mancha is a region

where many of the most important questions in European Bronze Age studies can be addressed.[13]

The La Mancha Bronze Age settlements date from about 2250 to 1500 BC – which fits in with the pattern of Minoan trading – with the Rio Tinto mines being worked for several hundred years, the miners later moving upriver and starting to farm on the Meseta, dotting the area with heavily fortified sites. As Dr Martin explains, the settlers in La Mancha coped with climatic uncertainties by practising relatively intensive agriculture.

The spectrographic analysis of bronzes from La Mancha sites show that they almost all consist of unalloyed copper of 96.9 per cent purity, with an arsenic content averaging 1.81 per cent. There is hardly any tin bronze. This arsenic content would have caused serious lung problems for the smelters.

Key to this argument is that ivory – which must have been sourced from Africa or India – has been found in excavation sites in the south of La Mancha. About 400 grams (14 ounces) of the material has emerged, mostly from buttons and bracelets: both raw, semi-worked up and as finished jewellery. It appears that ivory was worked up in La Mancha workshops – evidence of regular long-distance commerce. The well-defended, massively constructed settlements were also associated with new levels of intensive agriculture, of a more modern, Mediterranean character.

||||||||||||||||||

The early Iberian settlements with either mining operations or the structures to defend them were at Los Millares – copper; Almizaraque – silver; El Barranquete – gold; El Tarajal – both gold and silver; and Las Pillas – gold (see map). At that stage the work was primitive, the copper pure.

One particular mining culture, the 'Los Millares' people, caught my eye, thanks to the research of W. Sheppard Baird. Their name is derived from a major copper mining settlement of perhaps 1,000 people, on a site that had been discovered in the 1890s as the authorities were building a railway at Almería. Little is known about the Millares culture, but it seems to have spread across the southeast of Spain and it possibly reached as far west as the red-limbed Rio Tinto. As W. Sheppard Baird states:

> ... When they [the Minoans] surveyed the river basins of Almería in south eastern Spain they found everything they were looking for. For several centuries they probably would have been satisfied to sift the alluvial sediments for metals and established settlements in the river basin areas. Eventually they would have moved up to the inland sources of the alluvial metals to form permanent mining settlements, and that's exactly what they did. By 3200 BC many of the towns of the Aegean Minoan colony (Los Millares culture) had been founded.[14]

The intriguing thing from my point of view was that a wholly new influence penetrated the Iberian culture around 1800 BC. The Millares people suddenly took a giant leap forward in copper metallurgy. They began smelting arsenical copper: I would infer that the Millares learned so quickly at this early stage as a direct result of early contact with the Minoans.

Then, in turn, a mysterious new people, the El Argar, quite suddenly took over from the Millares. Excitingly, the El Argar influence began early, around 1800 BC, but it then developed: in 1500 BC the El Argar people entered their so-called Phase B, at which point a detectable Aegean presence infiltrated the culture. New introductions included huge *pithoi* – vases like the ones I'd first seen on Crete and Thera.

Crucially, from my perspective, the El Argar were people who had substantial links to the outside world. I am convinced that the El Argar were in truth the Minoans and that the vast expanse of southern and central Spain and its rich mining territory became one of their many colonies.

There are many clues to a history of Minoan colonisation at Los Millares – and to a later Mycenaean influence. For instance, its cemeteries have *tholos* (beehive) tombs identical to the Minoan beehive tombs that are found on both southern Crete and on the central Mesara plain. The remains of pottery, ivory and ostrich eggshells also hint at a culture that had substantial contact – at the very least – with our Mediterranean traders.

Ivory, bronze and ostrich eggs – suddenly, they all connected in my mind. Lisbon is possibly my favourite port in the world and I had been there many times. Had I not also heard tell of similar treasures, found in the ancient fortress of São Pedro at the estuary of the Tagus?

ıllllllllllllllllllll

Entering the Tagus estuary is an experience no sailor ever forgets – it's a truly magnificent natural harbour. The Tagus itself is a majestic river that knows no national boundaries. It strikes right through the heart of both Spain and Portugal, passing the magnificent, solitary splendour of Toledo, south of Madrid.

Vila Nova de São Pedro was a great Bronze Age fortress; built, archaeologists think, in a concentric style. It sat overlooking the Tagus estuary and was first excavated by the anthropologists H.N. Savory and Colonel do Paço.

To my delight, I discovered that there was a match, of a kind: São Pedro is completely contemporary with Los Millares. In his research papers, Savory draws parallels with the metalworking culture of the Los Millares late phase: 2430 BC. São Pedro's tombs did in fact contain exotic goods like ivory, alabaster and ostrich eggs – the classic signs of Minoan trading and influence. My memory had served me well. There was also evidence of a religious cult involving a female goddess – a deciding factor in Minoan religion.

Savory memorably describes the way in which c.2500 BC Vila Nova de São Pedro II had its own concentric fortifications. He makes analogies with Chalandria on Syros, another Minoan base in the Mediterranean, and his instinct seems to have been that the Early Bronze Age colonisation here came from the Mediterranean.

ıllllllllllllllllllll

I was hatching a plan. At some point I was due to give a lecture at Salamanca University. Marcella and I should take advantage of the trip and investigate the Tagus and its path through Spain.

The Guadiana River (see map) would take us to Évora. Just west of there, close to the copper mines at São Domingos, was my target, the Bronze Age astronomical site of Almendres Cromlech. A sister site of Stonehenge in England, Almendres was fascinating to me partly because it is not exactly unique. The devil, as ever, is in the detail.

Almendres is interesting in the context of this quest primarily because of one thing: it was built at the precise latitude where the moon's maximum meridian altitude is the same as the latitude of the site – 38 degrees 33 minutes north. This means that when the moon is at its highest, its orb is directly above the observer on the ground.

The only other latitudes where this happens are at Stonehenge and Callanish, on the west coast of the isle of Lewis in the Outer Hebrides.

Scholars already agree that the stones of Almendres, like those of Stonehenge, mark where the sun rises and sets at the equinox. The first excavator to investigate Almendres was the archaeologist Luis Siret. The evidence he saw convinced him that the new 'El Argar' settlers in Iberia were civilised seagoing traders who sought ores and kept the natives in the dark when it came to the huge value of the substances they traded. In his early reports Siret also mentions that the settlers traded in and manufactured oriental painted vases in red, black and green pigments. As we know, the Minoans' absolute speciality was colourful pottery. Black and green are colours derived from copper. The settlers also brought objects such as alabaster and marble cups, as well as Egyptian-type flasks, amber from the Baltic and jet from Britain.

So: we have strong evidence of a new culture arriving in Iberia, a people who were interested in the raw resources of gold, tin, copper and silver, and there is a gathering weight of evidence that these people were the Minoans.

This was the crucial link. The Minoans needed to read the stars: without that, they couldn't navigate. Astronomical stone circles could have been used to determine latitude and longitude and the dates of the equinoxes and to predict the positions of the sun and the moon and their eclipses way into the future. What if the Minoans who'd worked or supervised the copper mines at São Domingos and perhaps elsewhere in Spain had also – somehow – been involved with the astronomical stone circles at Almendres?

||||||||||||||||||||

For now, though, there were plans to make. We flew to Madrid, with the aim of heading through Ávila and Zamora to Lisbon. *En route* we would try to find evidence of Minoan visits following the course of the Tagus, either in museums or at archaeological sites. I rang one of the professors at Salamanca for advice and was advised to take a look at Almendres. I was also given an intriguing fact: that the La Mancha Bronze Age had ended abruptly. The well-defended, massively constructed settlements came to a relatively sharp end about 1500 BC – at the time when Minoan Crete was also abruptly crushed by the Theran tsunami.

The Morra, Motilla and Castillejo settlements of La Mancha were suddenly abandoned – virtually all of the radiocarbon dated sites withered within a century of one another. I would later find out that Bronze Age mining ceased equally suddenly in Britain, Ireland and America at around the same time:1500 BC.

21

SPAIN AND LA TAUROMAQUIA

But we still had some time to spare in Madrid. I asked the professor: what should we do?

'Oh, go to the Prado. It's unmissable.'

Unmissable it certainly was. The Museo del Prado has one of Europe's most extraordinary collections of art. The collection once belonged to the Spanish royal family. We went through gallery after gallery of the most exquisite work – from *The Garden of Earthly Delights* by Hieronymus Bosch to Rubens' *The Three Graces* – very robustly fat, as he chose to show them. We thought what we'd seen could scarcely be bettered until, with sore feet and a desperate need for something wet to keep the energy up, we stopped for a cup of tea.

Marcella had a teacake and a fresh injection of blood sugar. While I sat, still stupefied by the weight of my aching feet and the sheer enormity of the Spanish kings' echoing palace, she began diligently researching what remained to be seen.

'I'm not convinced I can take any more culture,' I grumbled.

Marcella firmly piloted me towards a quiet but large room, where a number of silent students, charcoal in hand, were absorbed in sketching. The gallery had a hushed and reverent atmosphere and I soon understood why. The room houses an electrifying series of sketches by the romantic Spanish genius Francisco José de Goya y Lucientes.

Goya's 19th-century series of etchings, 'La Tauromaquia', is a mesmerising sequence of images that document the bullfighting stunts and techniques used in his time. In his pictures, matadors stand on tables, or even chairs. Dogs are used to bait the bull. You see *'encierros'*, where the bulls are let loose to run through the town. In one picture the matador is standing firm ... while in another, he is pole-vaulting over the bull.

Sketch No. 90 is called: *The Agility and Audacity of Juanito Apiñani in [the ring] at Madrid*. The bold Apiñani is in the act of somersaulting backwards between the bull's horns, with the aid of a pole. The matador leaps over the bull's back, to land triumphantly behind its rear legs (see second colour plate section).

What was so familiar about this astonishing spectacle? It suddenly hit me where I'd seen it before – at the Minoan palace of Knossos. The fact that the image itself originally came from a miraculously long-lived fresco created sometime between the 17th and 15th centuries BC only made this all the more astonishing.

Was this a coincidence?

I thought not. Long ago in Medina, on the night before the Feast of the Assumption, I'd witnessed the extraordinary spectator sport of bull-leaping. Lately, I'd seen evidence of it again in Crete, southern India and Kerala. What I'd been witnessing – had I stopped to think about it – was a calling card. It had been sitting out the centuries, but it was there, written in the colourful script of the Minoans. This practice, I was sure, was a signature that the Minoans had left for us to read: those same Minoans who had probably first emerged from Çatalhöyük in southern Anatolia – possibly the world's first city and a place that the Minoans had perhaps first built. The evidence from Çatalhöyük shows that its people worshipped the bull; logic tells us that when they then reappeared in Crete the bull became central to their sacred festivals.

I'd found images showing the practice of bull-vaulting wherever I'd since speculated the Minoans had been; starting at Crete's palace of Knossos and voyaging all over the Near East to the Egyptian Nile Delta, then Syria, across the Aegean and all the way down to the south coast of Kerala in South India. The practice had also survived here in Spain, a popular inheritance that has lasted from the 3rd or early 2nd millennium BC to the present day.

The spreading cult of the bull was also backed up by a growing number of archaeological artefacts. Thanks to the 'Beyond Babylon' exhibition at the Metropolitan Museum of Art, I'd seen figures leaping the bull in Babylon, the image stamped clearly on a clay seal. In Athens' National Archaeological Museum is a bronze ring that shows a clean-shaven bull-leaper wearing a Minoan-style loincloth somersaulting on to the bull's back. His long hair flows in the air. Again, at Kahun in Egypt an image on a wooden box shows the leaper's epic dance against death. At Çorum in Turkey, a vase from an old Hittite settlement is decorated with thirteen figures gathered around the bull while once again the bull dancer plays out the dangerous game. In Antakya you will find a similar scene in a simple black and white drawing.

A distinct spring had come back into my step; all tiredness forgotten. What was not forgotten was an image that was still alive in my head. It was from the Archaeological Museum on Crete and to me it almost proved everything about my theory. Tiny, but telling: it was a Minoan medallion, about 6 centimetres (2.5 inches) wide, that I'd noticed in a cabinet towards the end of the rows of exhibits. It depicts a bull being led on to a ship.

'I've got gallery knees,' I announced, as we emerged into the wide avenues and blasting heat of Madrid. 'And museum legs.' But I wouldn't have missed the drawings I'd just seen for the world.

22

BLAZING THE TRAIL TO DOVER

Southern England: the 20th century AD. It was 1992 and workmen in Dover, England's busy passenger port, were digging an underpass to link with that wonder of modern transport technology, the Channel Tunnel.

As they dug down below what had been the debris of the medieval town – and then went even deeper, past the Roman layer – the men struck the perfectly preserved remains of a prehistoric ship. Downing tools, they stared in amazement. They knew, because of the depth at which they were working, that they were looking at something extraordinary. Gleaming in the narrow beams of their tungsten lights were the gnarled, blackened remains of a wooden ship.

Archaeologists mounted a rescue operation to record and salvage the timbers, cutting the boat into thirty-two pieces after carefully photographing and recording each piece. They were concerned that the ship, which had been preserved in the wet clay for thousands of years, would disintegrate into a puff of smoke once it reached dry air. So as they lifted the thirty-two sections by crane they placed them in a special container filled with preservative chemicals, itemising them meticulously so that they could be sure to replace the pieces according to the order in which they had been found.

This amazing rescue took just twenty-two days, a great tribute to all involved. Today the consensus is that the Dover Boat deserved all of its reverential treatment. The oaks that had made it had been cut down around 1500 BC. This boat was sailing long before Tutankhamun ruled in Egypt and at a time when ancient Britons were still using Stonehenge. Yet it is not just its great age that makes this ship remarkable. As one of the few pieces of complex technology from the Bronze Age that has survived almost intact, it tells us a great deal about the era; and Britain itself.

The boat is one of the oldest found in the world. What strikes the visitor instantly is the vessel's brute strength. Her timbers, although black with age, are perfectly preserved, glinting with what looks like a veneer of jet. The planks which form the hull were carved, rather than sawn, from whole tree trunks. It's what's called a 'sewn' boat; thongs bind the planks together, which were caulked with moss and beeswax.

The train from Victoria was running late that day and I could feel the anxiety of the other travellers as they thought about their delayed meetings, or missed connections at the port. Luckily, I didn't have any such time pressures. Looking at the soft-focus mist that was beginning to descend, I tried to project my thoughts back to the morning of 26 August 55 BC, when Julius Caesar arrived with his Roman invasion fleet. He came, he saw, but he didn't conquer. Nevertheless, by the 3rd century AD, when Dover's so-called 'Painted House' was built (alternatively known as the brothel), the town was a fully fledged Roman settlement. Even the ancient grass trackway, which became known as Watling Street, was paved over on its way to Canterbury.

So I paid the £2 ticket price for a round-trip bus ride and settled down to see 'Dubris', the town the Romans founded on the River Dour to defend their British interests. It was amazing to think how much must still be lying undiscovered underneath the ground, forever locked up by the mass of modern, Georgian or even medieval buildings that have been stuffed on top of Dover's Roman layer. No one had even known how much of it was here until the 1980s, when the town council dug up a car park and found a Roman fort. Up on the hill, Dover Castle looked down haughtily as I struggled to see the *Pharos*, or Roman lighthouse, through the gathering mist.

We got to the market square and I ambled off to see the boat itself, wondering what I would find. What I saw was, in some ways, a shock.

I thought back to the light-as-air ships of Thera's frescoes, greeted by diving swallows swooping around them; the crowds massing to see them as they arrived safe home from a major sea voyage. And then there was the steeplechaser that was the Uluburun. By comparison the British boat is an arthritic old carthorse. The Dover Boat had no rudder, mast or sails – all of the power came from rowers or paddlers. She also looked extremely heavy – more suited to the calm of a meandering river than the open sea. To view the Dover Boat please go to the second colour plate section.

And yet, says the ancient ship expert Peter Clark:

We have archaeological evidence that contact [with Europe] took place by the early Bronze Age, so that it is generally accepted that there must have existed a capability for cross channel voyages and coastwise travel along the Atlantic seaboard, extending from Iberia, round Brittany to western England ... the scene changes from the early Bronze Age onwards, with the discovery of the examples of plank-built boats found to be widely distributed round the British coasts. Some of these might reasonably be believed to have had the ability to travel coastwise, or when conditions were favourable to make short sea crossings.[15]

According to English Heritage, the boat's conservers:

... The Langdon Bay hoard of bronze implements, largely of French origin, found by divers just east of Dover harbour – clearly lost in the wreck of a sea-going vessel and of middle Bronze Age date – is good enough evidence that there was routine traffic between Britain and mainland Europe at this period. The Dover boat provides convincing evidence of an actual vessel with appropriate capability.

Of course I don't doubt for a minute that cross-Channel trade took place in the Bronze Age. Dover to Boulogne is the shortest stretch of the English Channel; and many historical sources mention various forms of commerce crossing the seas. Pytheas the Greek describes the Cornish tin trade in some detail. The first person to realise that the tides were caused by the moon, Pytheas made his journey to Britain in the 4th century BC – but judging by the way he discusses the trade, it is clear that it had already been in existence for centuries. Cornish tin would have been a valuable prize: it is cassiterite tin, which is harder than other forms and twice as shiny. Cassiterite still prompts wars in the Congo, where it is now found. Ancient Britain had so much of it that the Phoenicians named the country 'the Cassiterides'. [16]

There is much more evidence of the roaring trade in tin. Bronze Age specialist Professor Barry Cunliffe describes divers recovering forty tin ingots from a wreck found at the mouth of the River Erme as it enters Bigbury Bay. Found within 26 metres (85 feet) of each other, all of the tin ingots were quite clearly cargo from a vessel which had foundered – perhaps near Portland Bill, well known as the Cape Horn of the English Channel. Several pieces of ancient timber were recovered in the same area and were carbon dated to more than 4000 BC – 'far too early for the tin trade', according to Cunliffe. But is 4000 BC really far too early? The dates when we assume international maritime trade was taking place are continuously being pushed back.

Whether the Bronze Age came to Britain in 4000 BC or in 2300 BC – the generally accepted date – we can say for sure that boats like this could have been carrying tin from Britain to Europe by 1500 BC, the date of the Dover Boat. But this boat had nowhere near the sophistication, strength or resilience of the Mediterranean ships shown on the Thera frescoes. The Dover ship's weight and inflexibility also meant she could carry far less cargo than the Uluburun or Theran ships and so she must have been far less suited to ocean-going trade.

The English Channel tides are strong, which, coupled with the funnelling effect of the French coast, means that in any wind short steep waves form – they are maybe as high as 2 to 2.5 metres (6 to 8 feet). This is known as the 'Channel Chop', and I can't imagine that the Dover Boat would have fared well against it.

A few short calculations gave me a picture. I imagined the Dover ship being paddled by sixteen strong sailors for five hours at 2 knots across the English Channel, with 3 tons of copper or tin on board – while four men furiously bailed out. With that amount of physical blood, sweat and tears, they could only have rowed for five hours a day. A ship like the Uluburun could have kept going for days on end, with more than four times the cargo capacity on a like for like basis: and without leaking.

The Uluburun sailors would also have lowered the ship's centre of gravity and increased stability by placing copper and tin ingots as ballast at the bottom of the hold, beside the keel. This would have reduced roll, pitch and sagging. Placing ingots low in the hull of the Dover ship would not have had the same effect. The weight of the planks higher up her hull would have made her comparatively top-heavy.

When it came to trading, the Bronze Age Brits faced world-beating competition. Had the Minoans seized that opportunity by the horns of their deity, the bull?

||||||||||||||||||

I settled down to delicious bacon and eggs in a handy 'greasy spoon' cafe, nose in book. It felt good to be reading the work of a Roman naval commander and author in a Roman city like Dover. I was beginning to catch up with my neglected classical education, and I was having to do that fast. There is a well-known but apparently much disputed passage in Pliny the Elder's writings, that concerns the history of the British:

> Next to be considered are the characteristics of lead, which is of two kinds, black and white. The most valuable is the white; the Greeks call it 'Cassiteros' and there is a fabulous story of its being searched for and carried from *the Islands of Atlantis* [my italics] in barks covered with lead ...

> Certainly it is obtained in Lusitania [Iberia] and Gallaecia [Brittany] on the surface of the earth, from black coloured sand. It is discovered

195

by its great weight, and it is mixed with small pebbles in the dried bed of torrents. The miners wash those sands and that which settles out, they heat in the furnace.[17]

So: 'Atlantis' again, but this time the connection is being made by a Roman, not the Greeks. By the early years of the 1st century AD, Pliny seemed to have identified 'Atlantis' as both the past and the present source of the tin being shipped into the Mediterranean.

Minoan seafarers in search of those precious Bronze Age metals would have had to turn north towards Britain and cross the notorious Bay of Biscay in their markedly superior boats. One long 310-mile (500-kilometre) lee shore, the Bay stretches along the northern coast of Spain and around the western coast of France. The crews of the sailing ships of old had a great fear of being driven ashore here – or worse, wrecked – by a westerly gale. Biscay is notoriously stormy. Many mariners have run before the Atlantic's driving currents and ferocious gales, only to founder when they struck the Bay's vicious reefs and its shallower coastal shelf. June and July are the best months to cross: it is important for our seafarers to sail before the middle of August, to avoid the first autumn storms.

Even when sailing the relatively placid Mediterranean, it would have been foolhardy to travel before the beginning of May, after which the Mediterranean winter gales should have ceased. That should mean reaching the Straits around late August. Overwintering in Sanlúcar, the Minoans could have repaired any broken steering oars and sails and careened the ship to remove barnacles from the hull.

Let's imagine a sophisticated band of travellers, the Minoan seafarers. A hardy bunch of persuasive, charming and determined people, they are wily and well travelled. From what we'd seen on the frescoes, they are also lively company and extremely good-looking into the bargain. Their metallurgical know-how is unrivalled, in a society that values metals, and magic, above all things.

They have one goal in mind: riches. They need to find precious metals, make successful trade deals, and return to Phaestos, Knossos or Thera as rich men, bearing the raw materials that drove the most advanced technology of the entire age: copper and tin.

Each member of the crew is a hero. He has seen the wonders of the world, and will return to his people bearing the future in his hands – or within the hold of his ship.

23

THE LAND OF RUNNING SILVER

The early Greek historian Diodorus Siculus called Britain 'the bright country'. And indeed, Cornwall, the first part of Britain any seafarer striking from Portugal through France is likely to hit, does have a remarkable quality of light. But I think Diodorus' description is more likely to have been inspired by the sparkling richness of the country's metals. This was an island where the sun set in a blaze of tin, copper and gold.

After I'd gone to see the Dover Boat, and satisfied myself that foreign traders had little to fear from native competition, Marcella and I decided to make a week of it. We hired a cottage on the Cornish coast near St Mawes. From here we could explore the Carnon River, which in the Bronze Age almost ran with sparkling, silver-coloured tin ore.

For once we found that the subject was not controversial. English people in general, and Cornishmen in particular, readily accept that in the Bronze Age foreign sailors came from the eastern Mediterranean to mine tin. The usual date being given was around 2000 BC.

In fact, all sorts of folk memories have grown up around the export of the ore. The brasswork in King Solomon's Temple is said to have been made from Cornish tin, and an old legend has it that Christ himself visited Cornwall with his merchant uncle Joseph of Arimathea, who came to buy precious ores.

Much of Cornwall's bedrock is granite. As this cooled many millions of years ago, fissures and cracks opened up when the granite was still molten and hot rock from the earth's core bubbled up through the cracks. As these new rocks crystallised, they formed mineral lodes – tin, copper, zinc, lead and iron, with a little silver. Because the ore-bearing rocks came from vertical cracks, they had to be mined vertically – straight down into the earth.

Streams often ran across the tin deposits and sometimes sliced through them. Tin, being so heavy, was often left on the stream bed. In this part of Cornwall evidence of Bronze Age mining is everywhere, even to the body of a poor Bronze Age miner found at Perranarworthal.

The richest area of all in those days was the Fal Estuary (see map) and particularly the rivers to the west of it; notably the Carnon River, which was navigable throughout the Bronze Age, upstream as far as Twelveheads.

Another crucial factor that would have made Britain highly attractive to the Minoans was its incredible woodlands – as well as the type of wood that grew. From the air, Britain would have looked almost as the Amazon jungle does today: a mass of vital, living green, punctuated with the chattering rivers that shone with glittering grains of tin ore.

The mighty hornbeams, feathery oaks and majestic beeches and willows of Britain all respond well to pollarding and coppicing (cutting off young tree stems close to the ground). Much like a good haircut, cutting the wood short does two things. Firstly, it makes the tree healthier, so it lives longer. Secondly, it makes it grow more branches and produce more wood, a substance that is vital, as we already know, for the smelting process. Wood, as the *Epic of Gilgamesh* implies, was already becoming a scarce resource in the wider Mediterranean.

Why should we believe it was the Minoans, rather than the Phoenicians of a later time, who reached Britain and began to exploit its resources? Partly because of the great Akkadian Emperor Sargon I, 'the Magnificent', who lived around 2333–2279 BC. He commissioned a 'road-tablet', which recorded the mileage and geography of the roads through his vast Mesopotamian empire. A copy of it, made by an official scribe in 8 BC, was found at the Assyrian capital of Assur.

The tablet details 'The land of Gutium' and 'tin-land country, which lies beyond the upper sea (or Mediterranean) ... '.[18]

This latter reference was translated slightly differently by a former Oxford professor of Assyrian studies, Professor Sayce. His version reads: 'To the tin-land (Kuga-Ki) and Kaptara (Caphtor, Crete), countries beyond the upper sea (the Mediterranean).' In other words, his translation links the 'tin land' and ancient Crete together.[19]

There's another link in the evidence we should bear in mind. Why was there a sudden switch in Britain from copper axes to bronze ones between 2800 and 2500 BC? And why, around 2200 BC, did the tin content of British axes suddenly leap from virtually zero to 10 or 11 per cent? This date coincides neatly with the height of the Minoan trading empire, which I now propose was well under way. As Professor Cyrus Gordon wrote: 'The existence of an ancient formidable commercial network of which the Mediterranean Sea was the epicentre is being revealed.'[20] The Minoans had the technology: fast, sail-driven, ocean-going ships and the knowledge to navigate them. They also had the all-important know-how, when it came to smelting and processing their finds, and the advanced technical and design skills necessary to forge beautiful objects in metal. They had an insatiable thirst for bronze, the wondrous material that was helping their culture become the world's most advanced, in the process turning the Minoans of the ancient Mediterranean into some of the wealthiest and most powerful citizens of the known world.

The Minoans would have been drawn to Britain by the lure of rivers running with tin. They would have been unlikely to meet with much local competition. Imagine their jubilation when they discovered that another great prize was also here: copper.

24

A LABYRINTH IN DRAGON COUNTRY

Any ship sailing north from Cornwall's tin mines up the St George's Channel would have noticed a handsome group of mountain peaks near what we now call Caernarvon (Caernarfon), in northwest Wales. The curious mariner might then have turned eastwards into Colwyn Bay on reaching Carmel Head (see map), to investigate. This is what I believe happened in the Middle Bronze Age around 2500 BC, when the first Minoan explorers, having gathered a load of Cornish tin, pushed north to explore further.

Before the age of the dinosaurs this rocky foreshore was a tropical seabed. The huge carboniferous limestone headland of today began forming about 300 million years ago. Stone Age man would almost certainly have shared the Orme with mammoth, lions and the woolly rhinoceros, but archaeologists don't know for certain when *Homo sapiens* first arrived on the headland.

Many thousands of years later, two mining enthusiasts were investigating 19th-century mine workings. Wales is famous for mineral exploitation on an industrial scale in that century. But what Andy Lewis and Eric Roberts came across instead was a spectacular warren of copper mines that didn't conform to either 18th- or 19th-century mining practices.

Worried about exploring too far underground without taking safety precautions, they managed to get the local council to agree to do a survey. What that survey found shocked everyone, including themselves. A labyrinth.

Beneath their feet was a maze of prehistoric mine workings that extended into a dense warren of shafts, tunnels and side-chambers.

The dark, buried passages and chambers were immeasurably old, not at all like the shallow-cast pre-Victorian mining the Welshmen had been expecting. Quite by accident, Lewis and Roberts had unearthed the enormous Great Orme Bronze Age copper mine.

The Great Orme Exploration Society meets every Thursday night in the King's Head in Llandudno. So today you can visit them and drink a fine pint of bitter to Lewis and Roberts and those ancient pioneers who opened up the land, and the world, to a shining past: a technological age we had scarcely dreamed of.

I visited the site in late spring. It is an easy journey from London: first flat out to Warrington and then taking a leisurely local train westwards along the beautiful Flint and Denbigh coasts for another hour and a half. To the south lay the intriguing snow-capped peaks our Minoan mariners would have sighted; to the north a succession of shallow bays where they could have beached their craft. Today the coastline is a mixture of caravan sites, oil refineries and field upon field of Wales' ubiquitous sheep.

The terminus for my journey was the genteel yachting town of Llandudno. The Great Orme mine is halfway up Llandudno's 'peak', which is reached by a cable car and a tram that connects the town to the summit. It would be a spectacular and healthy walk up to the very top, where the Telegraph Inn was once the vital communications link between Holyhead and the busy global trading port of Liverpool, advising of the imminent arrival of sailing ships laden with valuable cargo. Still, I cheat and take the charming, stubby little blue-painted tram, which drops me off at an unassuming group of low, white-walled buildings – and puddles of grit and mud.

Inside the Great Orme mine one's first reaction is incredulity. If you were to draw a cut-through diagram of this place, its tunnels would look like the branches of a vast spreading oak tree, stretching right through the entire hill. The complex is different to anything I have ever seen – it's almost as if you are entering a huge subterranean sponge or ants' nest, where humans would have crawled down into the tunnels, crevices and fissures, steep shafts and side chambers. Some of the tunnels are so narrow that it seems that all those thousands of years ago only children could have excavated them. What a Herculean task it must have been.

Judging by the archaeological evidence, the first miners used antlers and shoulder blades to scoop out the ore before it was hauled to the surface by a system of ladders and windings. Once above ground, the copper ore was pulverised with stones. They were the shape and size of ostrich eggs. Usually coloured a rich cobalt blue, the ore was smashed out and then heated in a heaped crucible, using a pair of bellows surmounted by a pile of charcoal.

The liquid copper could then simply be poured into the mould as required – the process was as easy as that. When cooled, the copper could be worked in its own right, or it could be mixed with molten Cornish tin to create the all-important alloy of hard bronze. The end product – whether axe, knife or sword – was either hand-beaten or shaped by clay moulds. Sharp weapons and tools could be made stronger and more flexible by beating them after the initial moulding.

So here in north Wales we have the first industrial process in Britain, which started 4,500 years ago – a date determined by the carbon found in the charcoal residue here. Could this operation possibly have reached this level of sophistication so fast without the input of outsiders?

The origin of the word 'Orme' is lost to history, although we do know that many centuries later the Norsemen often used the term to refer to dragons, or 'worms'. From the sea, the great headland could indeed have looked like a sea serpent as it stretched its neck out into the waters. Perhaps the idea reflected exactly how much treasure lay guarded in the bowels of the earth.

For this was a hoard indeed. Mining engineers have calculated that 1,700 tons of copper must have been extracted from these ancient mines. That is enough to make more than ten million axes – three for each man, woman and child who then lived in Great Britain.

Alternatively, look at it this way. The copper could have supplied enough bronze for three million saws. Enough for the pyramid builders, 2,000 miles (3,210 kilometres) away in Saqqara, ancient Egypt.

TECHNIQUES WITH BRONZE AND COPPER

Professor R.F. Tylecote of Durham University describes the process of pouring copper or bronze into moulds to make bronze tools and weapons. He begins with the basic problems of casting. First, when metals are heated in solid fuel furnaces they absorb gases from the fuel – wood, in the case of the Bronze Age. Of these gases, by far the most troublesome is water vapour, which dissolves in contact with the metal into copper oxide and hydrogen. The hydrogen enters the liquid metal and stays dissolved until the metal cools and begins to solidify, when it emerges as gas bubbles which spoil the casting.

This problem can be alleviated by giving the gas plenty of time to rise to the top of the molten metal – that is by cooling the liquid metal slowly. That is easier if you are making a large amount of alloyed metal at one time.

Another problem is shrinkage, which occurs as the metal cools and contracts, causing cracks and cavities in the casting. These cracks can be filled by pouring in more liquid metal from the feeder. In a thin section like a sword or a socketed axe, a washer was removed and the sides of the casting mould were moved slightly together as the metal cooled. The majority of Bronze Age castings were of this type. The density of the cast object can be increased, and cracks reduced, by hammering the metal after the casting – this was done in the case of flat copper axes, principally to harden their cutting edges.

The better the mould, the sharper the axe. By the early Bronze Age, stone moulds were used. There was little variation in this type of manufacture. Moulds were cut from blocks of stone, with cavities for two or more axes, the moulds being closed together with a flat removable washer or gasket between the two. That would be withdrawn as the metal started to cool. In the early Bronze Age, most of the stone for the moulds came from the Pennines, a mountain chain in the north of England.[21]

The whole experience resonates in my mind as I return to the hotel for a quick wash and brush-up, followed by a snifter at the bar. A fascinating picture of prehistoric Britain is coming to life in my mind: a society with a level of industrialisation and organisation that is astounding. This is not anything like how we tend to think of ancient Britain: perhaps we need to think again.

The copper ore is mined in Wales, the tin in Cornwall. The axes are made in north Wales from smelted metal, using local charcoal. Casting is done with hard limestone from the north of England. The large numbers of new bronze-workers eat locally farmed Welsh cattle and lamb; clothiers and shoemakers use the hides. All of these workers need bread and the best corn lands are in the dry, sunny east of England. Woods and forests there are cut down, using brand new Welsh axes. The grain is then transported by ships constructed from crude planks made from the swaying trees of eastern England – which are taller and thicker than wind-blown Welsh trees.

At the mine itself, the dendrochronology (tree-ring dating) of the charcoal shows that the miners used standard-sized, pollarded branches. So we now have yet more groups of workers at Great Orme – not just miners, but foresters, charcoal-makers and a host of other tradespeople; huge numbers must have been involved. The miners used antlers and bones for their digging – these animals had to be hunted. Their skins were stretched and dried to make leather clothes and shoes – which means leather-workers, clothiers and shoesmiths. More and more men kept their hair tidy and shaved with bronze razors; they also used copper eyelets to tie up their clothes and button their shoes. Copper from the mine then began to be hammered into decorative ornaments and jewellery, as locals became more wealthy.

For the Minoans and their budding empire, Britain would have been like a treasure trove. It was a treasure trove, moreover, that gave them a jumping-off point for the whole of northern Europe. Perhaps even further.

That tell-tale prickling of excitement was in my veins again. I was going to track my Bronze Age pioneers even further, if I could. Another sea voyage of my own beckoned.

25

STRANGE BEASTS AND ASTROLABES

The dead of night: Saxony-Anhalt, northern Germany: 1999. Two hundred miles (322 kilometres) away from here, at the amber-trading city of Rungholt, is the very place where Hans Peter Duerr had found his Bronze Age Minoan cooking pots and had concluded that the Minoans had been sailing here – routinely.

Three black-clad figures were combing through a deep forest at Nebra, with illicit metal detectors.

After several cold, dark hours of scanning the forest floor, the men found themselves in a small clearing near a hill. Suddenly their detectors came alive. That high-pitched whine meant metal – lots of it. They tore into the earth with pickaxes. After a brief struggle that earth gave way – and yielded up a treasure it had kept safe for more than 3,000 years.

Carefully, one of the men picked a strange, flat object out of the hole and delicately brushed it free of the clinging forest loam. What was it? Pocked and covered with a green-bronze sheen, in his hands lay something that looked strange, almost magical.

A bronze disc 30 centimetres across; its surface inlaid with gold. Even in the moonlight, the men could see that they had found something special; something that drew an incredibly vivid image of the heavens. You can view the disc in the second colour plate section.

The sun – or possibly a full moon – faces a crescent moon. The two images are divided by what seem to be stars. The surface of both the sun and the moon are pockmarked with metal corrosion, making them look eerily realistic as if their real, crater-ridden counterparts had been studied through a telescope. A boat navigates the sea beneath. To a traveller the boat design seems similar to the barge of the Egyptian sun god, Ra.

This object is unique. Nothing like it has ever been seen before. We now know it as the 'Nebra Sky Disk'.

A few years later, the Nebra Disk was rumoured to be circulating the black market with a price-tag of a quarter of a million pounds. Unauthorised digging of archaeological finds is a crime in Germany. In an elaborate sting operation dreamt up by Harald Meller, who had just been appointed head archaeologist at the nearby museum of Halle, the looters were caught in a hotel – and the Disk was rescued.

Meller was the hero of the hour. He had won the Disk for posterity – and for his museum. When he finally got his find back to the calm and quiet of his office, he must have smiled in relief, and then thought: 'After that huge struggle to get hold of it – now what?'

What exactly had they won? Was it an astronomical device? The surface of the mysterious Disk was a mass of symbols. Were these just random images, or did they mean something more? The astronomer Professor Wolfhard Schlosser, of the University of the Ruhr, was called in to try to verify if these symbols might in fact represent heavenly bodies – the constellations.

Could northern Europeans have been advanced enough in the Bronze Age to have mapped out the stars? If not, the Disk's existence supported my theory that the Minoans had been here – and that they had had access to knowledge at least the equal of that of the Babylonians. Professor Schlosser's first move was to isolate the largest group of 'stars'.

The 'star' marks were spread out in a pattern across the object's surface. The professor ran them against a recognised computer programme to see if they would match with the stars in the night sky, first in the northern hemisphere and then in the southern. But there were no matches. These dots, it seemed, were just decoration.

Then the professor looked at the small cluster of seven stars in the middle of the Disk, right between the circles that might represent the sun and the moon. They seemed to form a distinct pattern. Could that be a constellation?

Professor Schlosser quickly realised that the cluster resembled one above all others: the Pleiades. The ancients thought there were only seven stars in the cluster. To them it was one of the most beautiful in the night sky. Significantly, in the mythology of the ancient world, where the greats literally became stars, it was celebrated for being made up of the seven daughters of one of the great Titans. His name was Atlas.

Today we know that the Pleiades is made up of eleven main stars, but only some of them are easily visible to the naked eye. So Schlosser turned to the oldest images of the Pleiades that he could find: tablets and scrolls from the East. And there he saw a wonder: the Pleiades, drawn with just seven stars. An image just like that on the Disk.[22]

The Nebra Disk had been found buried with a trove of bronze artefacts: two bronze swords, two hatchets, a chisel and fragments of twisted metal bracelets. Although the swords themselves look to be of German design, their metal content is not. Analysis found that the gold was from the Carnon River, where Marcella and I had been staying in Cornwall. The tin content of the bronze was also from Cornwall. If the swords were in fact made at the same time, they dated the Disk to c.1600 BC.

Two golden arcs run along the sides of the Nebra Disk: they appear to have been added later. Those mysterious arcs tell another story, but to relate it we need to travel a few miles further, to the Saale valley.

High on a plateau overlooking the valley, just 15 miles (25 kilometres) away from where the Nebra Disk was first found, is the tiny hamlet of Goseck. Here is another hidden treasure, one which was only discovered by chance in 1991, by aerial reconnaissance. A large, double concentric ring of post holes, pierced by gates and surrounded by a circular ditch. Again, as a specialist in astro-archaeology, Wolfhard Schlosser was called in to investigate.

The first clue as to the function of this new discovery was the fact that the gate leading into the Goseck circle's wooden palisade is precisely aligned with north. The site was positioned to observe the movements of the sun, the moon and the stars and for keeping track of time. The southern gates marked the sunrise and sunset of the winter and summer solstices.

Schlosser believes that the Nebra Disk and the circle are connected and that the constellation patterns on the Disk were based on previous astrological observations, possibly made over a period of time at Goseck. The two golden arcs that were added to the Disk, he reasoned, must also mark the winter and summer solstices. Spanning an angle of 82 degrees, the arcs mark the same angle that occurs between the positions of sunset at the summer and winter solstice at the latitude of this area, Mittelberg (51.3 degrees north).

I think that the Nebra Disk is a device that links the sun's movement with that of the moon. What use would the people here in Saxony – whoever they were, whether locals or visiting travellers – have put that knowledge to? Well, the very name 'the Pleiades' is from the same root as the Greek word *pleio*: 'to sail'. We now know that in ancient times, the Pleiades' heliacal rising was used to predict the time that ships could set sail from the Mediterranean: from early May to early November.[23]

Needless to say, I believe that the device must have been brought here by the only people who had this sophisticated understanding of the skies, the Minoans. Hans Peter Duerr's experience is proof of the Bronze Age Minoans' journeys to this area. The only people who had the reason to be here – in this case probably bartering their precious bronze for the sumptuous prize of Baltic amber – were the Minoans.

What reason could they have had to create the Nebra Disk? Why, for example, would it have been found so close to an ancient wooden circle aligned with the stars? I am convinced that this is no accident. This beautiful little object, like the extraordinarily intricate gold seals found on Crete, has in fact a highly practical purpose.

I am sure the Disk was used, perhaps in ceremonial fashion, as an aid in navigation.

There was now a huge, and to tell the truth somewhat overwhelming, question hovering in my mind. Almendres; the circle here at Goseck; the one near the Red Sea; the one in Kerala; and even Stonehenge. Why did stone and wood henges keep cropping up on my trail? I could no longer avoid the moment. It was now imperative for me to explore one of the world's most ancient mysteries: the ceremonial stone circle.

Notes to Book III

1. Philip P. Betancourt, *The Chrysokamino Metallurgy Workshop and its Territory*, A.S.C.S.A, 2006

2. Gerald Cadogan, *Palaces of Minoan Crete*, Routledge, 1991

3. K. Aslihan Yener. 'An Early Bronze Age Age Tin Production Site at Goltepe, Turkey', The Oriental Institute and the Department of Near Eastern Languages and Civilizations, University of Chicago, 2007

4. Richard Cowen, UC Davis

5. C. H. W. Johns, *Babylonian and Assyrian Laws, Contracts and Letters*, 1904. Project Gutenberg (www.gutenberg.org/ebooks/28674)

6. Richard Cowen, UC Davis

7. Theodore A. Wertime. 'Man's First Encounters with Metallurgy' in *Science* 25, December 1964, vol. 146, no. 3652, p. 1664

8. Oliver Rackham, *The Illustrated History of the Countryside*, J. M. Dent, 1996

9. Rodney Castleden, *Minoans*

10. The Thera Foundation (www. therafoundation.org)

11. F. Nocete, 'The smelting quarter of Valencia de la Concepción (Seville, Spain): the specialised copper industry in a political centre of the Guadalquivir Valley during the Third millenium B.C. (2750–2500 B.C.)', *Journal of Archaeological Science*, 35:3

12. Mark A., Hunt Ortiz, *Prehistoric Mining and Metallurgy in South West Iberian Peninsula*, Archaeopress, 2003

13. Concepcion Martin et al. 'The Bronze Age of La Mancha' JSTOR

14. W. Sheppard Baird, 2007 (www.minoanatlantis.com)

15. Edward Wright, in *The Dover Boat*, ed. Peter Clarke, English Heritage, 2004, p. 261

16. Barry Cunliffe, *The Extraordinary Voyage of Pytheas the Greek – The Man Who Discovered Britain*, Walker & Company, 2002

17. Pliny XXXIV, 47, Harvard Classics

18. In *Keilschrifttexte aus Assur verschiedenen inhalts* 1920, no. 92, trans. Professor Waddell

19. A. Sayce, *The Religions of Ancient Egypt and Babylon*, 1902, p. 3. Project Gutenberg (www.gutenberg.org/ebooks/35856)

20. Cyrus Gordon, *Before Columbus; Links Between the Old World and Ancient America*, Crown Publishers, 1971, p. 81

21. R.F. Tylecote, *The Prehistory of Metallurgy in the British Isles*, Institute of Metals, 1986

22. Schlosser, W. (2002), 'Sur astronomischen Deutung der Himmelsschiebe von Nebra', *Archäologic in Saschsen-Anhalt* 1/02: 21–30.//E. and C-H Pernicka, 'Naturwissenschaftliche Untersuchungen an den Funden von Nebra', *Archäologic in Saschsen-Anhalt* 1/02: 24–29

23. Theophrastus of Eresus, *On Weather Signs*, Brill, 2006 pp. 29, 43

BOOK IV

||||||||||||||||||||||||||||||||||||

EXAMINING THE HEAVENS

26

SEEING THE SKIES IN STONE . . .

A visit once again to Egypt: the majestic and formidable Minoan ally, and a cultural lynchpin of the era. Egypt was my next logical step in following the trail of ancient knowledge around the world. This is because the oldest stone circle on the globe lies on the Upper Nile.

The stone circle at Nabta was begun in the 5th millennium BC. I sensed that as privileged guests in Egypt, the Minoans might well have been able to study the Nabta site. I knew little about Europe's oldest astronomical observatory at Gosek, 200 miles (322 kilometres) up the Elbe, and the Nebra Disk found so close to it. Neither could I rely on written research, or much in the way of expert help: Gosek has been little studied. But Gosek and the Nebra Disk seemed to be pushing me towards a solution that was obvious enough to be staring me in the face.

A pattern was emerging. Wherever the Minoans had travelled, a stone or wooden circle seems to have appeared. Was this idea too improbable? I decided to follow my instincts and research the oldest stone circle in the world, to uncover the truth.

||||||||||||||||||||

I am convinced that the Minoans' drive to travel was supported by their astounding grasp of navigation. But to navigate you need reliable calculations about the stars – and that information needs to relate to your precise location on the globe. How did they obtain such information? Like many others, I strongly suspected that stone circles were built for astronomical as well as ceremonial purposes and that they were used for far more than predicting the seasons. I was convinced that the Minoans needed to build on the astronomical and navigational knowledge they had already gained from Babylon. The only way they could do that was to create their own observatories by building, or perhaps adapting, suitable structures. Stone circles.

At the edge of the Western Desert, 500 miles (800 kilometres) south of Cairo, near the border between modern Egypt and Sudan, lies the flat, arid bed of an ancient dried-up lake. It is a desolate spot between the desert springs of Bir Kiseiba and the shores of Lake Nasser, and about as far away from civilisation as you will ever find. Dust is everywhere: in summer the wind roars over the nearby sandy ridge. Once this place was green and lush, a seasonal lake filled by the summer rains. Now the land is a barren sea of sand, and the heat is crushing.

A millennium before the beginning of the Egyptian 1st Dynasty, nomadic tribespeople – cattle herders who normally roamed wide over the Sahara with their livestock – would congregate here with the arrival of those all-important rains. They slaughtered some of their precious cattle as a sacrifice of thanks.

Nabta Playa, as it is now known, has forced Egyptologists to rethink their theories about Egypt's origins. It was an enigma: a totally unexpected stone circle set in one of the world's most isolated spots, until a trunk road was put in to allow construction traffic to reach Egypt's New Valley Project. About 62 miles (100 kilometres) west of here are the colossal Nubian rock monuments of Abu Simbel, imposing huge sculpted figures that are now part of a World Heritage Site. The impressive statues were relocated by the authorities in the 1960s, to accommodate a new dam on the Nile.

Nabta, on the other hand, has remained exactly where it stands today. It seems as old as the sky. As Toby Wilkinson writes, all the signs are that the first peoples who lived here c.7000–6000 BC were much more sophisticated than their contemporaries in the Nile valley. They built both above and below ground, they had planned

216

settlements and they may even have imported their livestock from southeast Asia.[1] Greater Egypt would of course later catch them up.

Scattered through the moon-like landscape at Nabta are carefully placed stone megaliths: sentinels standing guard on the horizon. An oval ring of strange, humpback stones surrounds a group of uprights, mostly at differing heights. At the centre, two pairs of stones point north–south. Another pair is also pointing towards the midsummer sunrise. Why are the stones here? The answer is unavoidable, once you look at a map. The stone circle, built in the 5th millennium BC, is positioned exactly on the Tropic of Cancer.

The Tropic of Cancer is a latitude line that circles the earth. On this latitude, every year at the time of the midsummer solstice, the noon sun at its zenith hangs directly overhead. In other words, this is a very special place on the planet, a fact that the ancient Egyptians must have understood perfectly well. As at the far more rudimentary Goseck, the only logical conclusion is that Nabta must have been an astronomical observatory, that every June was used to prepare for the rains and watch the stars – perhaps even use them as a guide for global voyages.

Nabta's stone circle is far smaller than later observatories like Stonehenge. The first phase of development here began around 4800 BC. Later, between 4500 to 3600 BC, megaliths were dragged into new places – and aligned with Sirius, Arcturus, Alpha Centauri and the Belt of Orion. As the skies changed, more stelae were rearranged to align with the brightest stars like Kochab, in the constellation known as Ursa Minor: the Little Bear.

The 12-foot circle of stone holds four pairs of taller stones aligned opposite one another. As the sun rose on the summer solstice, when it was at its furthest apparent northerly position, the great burning orb would appear like a sudden omen through the two sets of standing sentinel stones. In other words, the stones are a window into time, marking the passage of the seasons. There were two sets of alignments – cardinal (north–south) and solar. This was all that was needed to mark the passage of the year in Bronze Age Egypt.

Because the circle is set exactly on the Tropic of Cancer, there would be no shadow cast by the stones at midday on the summer solstice. Moreover, this would happen at a time when day and night were the same length: twelve hours. The astronomers at Nabta must have realised that this phenomenon was caused by the earth rotating on its axis once every twenty-four hours, and on a repeating annual cycle. They would also have noticed that at sunset each day after the midsummer solstice, a different star to that of the day before rose on the eastern horizon.

In short, using this observatory you could work out that the sun, the earth and the stars were governed by different rules – the earth rotated once every twenty-four hours, while the stars moved independently of the earth and the sun.

Observers watching the stones' shadows would have noted the change as those shadows grew longer and longer throughout the six months that preceded midwinter. Then they started to shorten again until midsummer, when they disappeared. As the shadows lengthened, the sun grew cooler – because it was further away. Yet at sunset, different stars still appeared in the east each day. This would have led to the realisation that the earth and the sun were nearest to each other at midsummer and furthest away at midwinter. The observers would have worked out that midsummer appeared after 366 sunrises and the earth rotated every twenty-four hours, and would have deduced that either the earth circled the sun or the sun circled the earth.

This scientific examination of the heavens told the people the correct times to sow crops – when the Nile flood was nearly due – and when to harvest, when the floods should have peaked and passed. The Egyptian farmer, the *fellahin*, could merely plant and wait. Surely this shows that in 4500 BC Egyptian astronomy was among the world's most advanced? The Egyptians knew they could establish true north by examining a star within the Little Bear constellation, Ursa Minor. It was called Kochab, and it was then the pole star. We know they understood this because the Giza pyramids were also aligned with Kochab.

The astrophysicist Thomas G. Brophy suggests that the prehistoric stargazers who built the Nabta stone circle must have known a great deal more about the heavens than we assume. One of the stone 'doorways', he points out, is aligned north–south – this seems reasonable, because it aligns with Kochab's position. Brophy further suggests that the six central stones inside the circle represent the three stars of Orion's belt (the southerly line), while the northerly line of three stones stands for the three stars that defined the shoulders and head of Orion, as they then appeared in the night sky. These correspondences were for two dates, c.4800 BC and at precessional opposition.

In short, if Brophy is correct, the Egyptians must have worked out the long-term precessional 26,000-year pattern caused by the wobble in the earth's axis by 4800 BC. This change is what makes the night skies look different over time – and is the reason why we don't see the exact same patterns in the stars as the ancients saw. (It is also why all astrology, which fails to take account of precession, is bunk.)

Over vast periods of time – 26,000 years, to be precise – the night skies transform themselves into bewitching new patterns. This happens because the earth is slightly fatter at the equator: as it rotates it acts like a spinning top, changing posture in its ever-so-slow ellipse around the sun. The dance takes 13,000 years to complete and as our own position changes, so does the position of the stars we see. Today, true north is determined by the star Polaris. In another 13,000 years this inevitable process will bring a new star into the position of true north. Then it will be the bright little Vega, in the constellation Lyra. In 26,000 years' time, the cycle will be over and Polaris will once again be the North Star. It's a humbling realisation. This ancient site in the Nubian desert tells the story of a 'primitive' people who knew far more about the earth than the average university student does today.

Nabta's sacred landscape is also dotted with peculiar mounds, marked out with round, flat stones. Under one of them, in an underground chamber, a huge sandstone monolith was found. It is possibly the first monumental sculpture ever made in Egypt. The stone is carefully shaped and dressed to look like a wild beast: the unmistakable figure of a bull. I can't help but make a mental connection with the Minoans.

Their status as privileged guests in Egypt – as we know from the royal palace at Tell el Dab'a – would have meant that if they did not already have this astrological knowledge they would have been privy to the Egyptians' knowledge of the heavens. Moreover the Egyptians were in their debt, because they were reliant on the Minoans for their bronze and their tools.

By the time the Minoans reached the Upper Nile in King Amenemhat II's reign (1919–1885 BC) they could also have acquired much Babylonian astronomical knowledge, and might have been in a position to trade bronze finished goods with the Egyptians in return for that knowledge.

IIIIIIIIIIIIIIIIII

I was beginning to suspect that the Minoans used their understanding of Nabta to alter and improve upon other rudimentary stone circles they found on their voyages. These already existed, but had probably been built for different purposes.

There is an unmistakable pattern of observatories being found near mineral mines around both the Mediterranean and the Atlantic coastlines (see map). All were built between 4000 BC and 2500 BC:

- *Malta*
- *Sicily*
- *Portugal*
- *Brittany*
- *Ireland*
- *Britain*
- *Hebrides*
- *Orkney Islands*

All of these stone observatories were based on the same principles – and were built using the same system of measurements. They also had the same goals: to record astronomical events such as sunrise and sunset at equinoxes or solstices, the moon's meridian passage, solar and lunar eclipses and occasionally the rising and setting of Venus.

In order to reach Nabta, the world's very first stone observatory, the Minoans would have had to sail up the Nile: a round trip of nearly 1,000 miles (1,600 kilometres). Would this have been feasible? We know from Egyptian records that the pharaohs of the Middle Kingdom built locks to tame the cataracts and rapids of Aswan, so the journey to Nabta by river could have been possible. We can say for sure that the Minoans reached the Valley of the Kings, which lies three-quarters of the way to Nabta.

We know this because there are numerous Egyptian records describing Cretans bringing gifts to the pharaohs, whose court was then in the Valley of the Kings. Further evidence that the Minoans had travelled as far as Luxor, ancient Thebes, had come in the form of my old friend the American tobacco beetle, which has also been found at Luxor.

So Marcella and I put the idea to the test by hiring a felucca, and sailing upstream. Egyptian murals suggest that the basic rig of this traditional wooden sailing boat has been in use for thousands of years. The Nile was placid, running northwards towards the Mediterranean at a speed of about half a knot. A pleasant wind blew over from the Mediterranean, carrying us upstream at about 3 knots against the current. At that rate I reckoned the journey to Abu Simbel would have taken about six weeks, sailing eight hours per day.

There is some tantalising evidence that the seafarers had gone even further. Millennia before us, the Minoans seem to have broken their journey: at Tod. Here, underneath a temple dedicated to the hawk-faced god of war, Montu, French archaeologists had made a fascinating discovery. A treasure trove (see map).

This find fleshed out the bones of my theory, that the Minoans had travelled far into Egypt's vast interior. Tod was a huge surprise to me. The temple's stubbiness came as quite a jolt: unlike Egypt's more famous temples and pyramids it is a workaday building, much less monumental than I'd expected. There is a chapel there they call the 'birth house', dedicated to the female gods.

On the western side of the site is a well-preserved quay with paved flooring. It led to what was once an avenue of sphinxes and the main part of the temple. Was I treading on the very paths the Minoan explorers had once followed?

I was certain they had once travelled in this region. What the French archaeologist Fernand Bisson de la Roque unearthed in the form of the extraordinary Tod treasure goes a long way to prove it. Buried here at Tod beneath the floor of the temple of Sesostris I (c.1934–1898 BC) were four copper chests. They bear the cartouche of Amenemhat II (1919–1885 BC) of the 12th Dynasty – a pharaoh who lived at the height of Minoan trading power and influence.

The exotic hoard of treasure found in the chests may well have been a sacred offering to Montu, the god who is said to have slain the sun's enemies from the prow of a boat. It is really intriguing that an offering of Minoan precious gifts was made to Montu, since at this point in history the god was portrayed as having a bull's head.

Now divided between the Louvre in Paris and the Cairo Museum, the treasure had with it pieces of silver and gold ware that were clearly not Egyptian. The smaller chests contained silver cups with a design similar to ceramics from the Protopalatial period – c.1900–1700 BC – at Knossos.[2] The handle of one silver cup is the same as that on Minoan vases from the Middle Minoan period.

Other objects in the hoard seem to be mainly from the Levant and Anatolia – places that I now knew were regularly visited by Minoan traders. The necklaces are distinctly Minoan in style. In summary: if the Minoans had managed to reach the Valley of the Kings by the Middle Bronze Age, Nabta would have been an obvious next step. The Minoans took knowledge just as much as they took precious lapis lazuli and silver. By the same token, they sought it.

Sure that I was getting ever closer in the hunt for the truth, I took leave of Egypt and Nabta's eye-opening standing stones. The lead I needed to chase down now was something much, much closer to home: that is, Britain. My next stop would be the stone circles of Europe.

Yet my trip to Egypt had yielded another avenue of enquiry. If we take a leap of faith and assume for a minute that, as in Egypt, Minoan traders brought with them hoards of bronze then we should also be able to trace their progress as they moved north into Spain, then into northwest France and Britain. I needed to look at what they had left behind them – hidden underground.

27

MEDITERRANEAN AND
ATLANTIC MEGALITHS

As the far-sighted physicist Sir Isaac Newton once said, no great discovery was ever made without first taking a bold guess. Many academics have taken this to heart in examining the stone circles of Europe and beyond. They wanted to see if there was any common factor that explained why they were all such similar structures. My guess is that it was because they all share a great secret: the influence of the Minoans.

As the Minoans expanded their trading empire across the Mediterranean from Crete, first to the copper and tin mines of Iberia and then on to northwest France, Britain and Ireland, I believe they built – or more probably modified – circular observatories based on the Nabta blueprint they had studied in Egypt. A quick summary of these stone circles includes:

MALTA

Like Crete, Malta has a superb strategic position midway between the toe of Italy and Africa to the south. It also happens to lie midway between the copper and tin mines of Iberia in the west and the rich market for bronze to the east. Malta has been fought over by Arabs and Christians, the French and the British and finally the British and the Germans in the Second World War. The island finally won independence in 1964.

The archaeological record reveals unequivocally that around 2500 BC a new people carrying an entirely different culture arrived on the island. The new inhabitants disposed of their dead by cremation and made use of bronze tools and weapons. Both factors reveal their kinship with the warlike Bronze Age cultures occupying Greece, southern Italy and Sicily at around the same time. This is a strong reminder of Plato's story of Atlas and his brothers – brothers who were given kingdoms of their own.

A substantial stone circle appeared at Xaghra, a village on the smaller of Malta's two islands, with a megalithic structure inside it. It is uncertain whether there was one structure here or two, but it was here. And the timescale fits in with the pattern of Minoan bronze-bringing – and the Minoans' voyages of discovery.

SICILY – MEGALITHS OF MONTALBANO ELICONA

During my trip to Malta I was told Sicily had identical stone circles. We took a ferry from Valletta to Syracuse, where I had a serious disagreement with a taxi driver, who charged the equivalent of 70 US dollars for a two-minute ride from the ferry to the hotel. The taxi driver had powerful friends, so I ended up in prison for failing to pay more than a reasonable fare and languished there for the night. I was released at dawn and we set off.

The huge stones in the wild, romantic landscape at Montalbano Elicona are now interspersed with wild flowers and wind-bent yew trees. They were reputedly erected by 3000 BC and the stones aligned to the summer solstice. Some work has been done towards checking the alignments of these stones with other stars in the heavens, but, as at Malta, Sicily's famous site would reward more archaeological study. In form, Montalbano Elicona resembles a much smaller Stonehenge: a prototype, if you will.

CROMELEQUE DOS ALMENDRES, PORTUGAL

The Almendres stone circle we have met before. It was built on top of a hill 10 miles (16 kilometres) west of the Bronze Age copper mines of São Domingos (chapter 20) in around 4000 BC. The site consists of two circles, built in sequence. The result is an oval of ninety-two upright stones measuring 30 metres by 60 metres (98 by 197 feet). Some of the stones have decorative cut marks, spirals and circles, and there are stones that point to sunrise and sunset at the equinoxes. More interesting is the latitude of the site: 38 degrees 33 minutes north.

At this precise latitude, the moon's maximum altitude at its meridian passage is directly overhead. If you looked into a well, you would notice that your head was directly in the shadow of the moon. This is because the moon's orbit around the earth is in a different plane to the earth's orbit around the sun. As I've already mentioned, the only other latitudes where this occurs are at 51 degrees 10 minutes north, the latitude of Stonehenge in southern England and Callanish in the Outer Hebrides. This cannot be a coincidence – these three sites in particular must have been built by people who had the same astronomical knowledge. I would specu- late that this interest in the meridian passage of the moon may have been for religious reasons.

Luis Siret's finds at Almendres, especially the ceramic wares, bear the distinctive imprint of Minoan pioneers: how involved had they been here in Portugal?

BRITTANY

On the islet of Er Lannic in the Gulf of Morbihan in northwest France are two half-submerged stone circles. Both circles contain sixty stones, although only the northern one is still visible. The site was excavated in the 1920s by Zacharie Le Rouzic, who calculated Er Lannic had been erected c.3000 BC. Le Rouzic found that the lines of the stone circle pointed to the cardinal points of north, south, east and west. The site is conveniently near prehistoric tin and gold mines.

IRELAND

Irish stone circles of the Early Bronze Age are much smaller than their British counterparts such as Stonehenge, Avebury or Callanish, but nevertheless have stones which point to sunrise at the summer solstice and sunset at the winter solstice. Because Ireland has so many stone circles it was thought they were built for religious reasons. Recently, expert opinion has shifted, with many agreeing that they must have been created to study the skies.

NORTH GERMANY

I have already referred to the Goseck observatory on the River Elbe and the Nebra 'Sky Disk' that was found nearby. Goseck's draw for the Minoans would have been the amber trade, which will be described in some detail later.

OUTER HEBRIDES – CALLANISH

This observatory is one of the most interesting (see Almendres, above). It too will be analysed in more detail in another chapter.

This is also an opportunity to examine the claim of L. Augustine Waddell, that the people who built the circular stone observatories across the western world were the same people who mined copper and tin; and in particular that these stone observatories were built in places where the Minoans traded extensively. The Minoans seem to have adapted existing circles, but to have used stones in place of wood. They were able to do this because after 2200 BC they had the technology needed – bronze axes and saws sharp enough to trim stone.

||||||||||||||||||||

The circles I am interested in are in different parts of the western world and may even lead ever further west again – to North America. Those I have so far mentioned may have varied in shape and size, but they had many important things in common. They were often accompanied by *cursi*, ceremonial pathways marked out by megaliths, that led the way into the stone circle. Built using a common measurement, the megalithic foot, they were constructed to record the same astronomical events – usually the rising or setting of the sun at equinoxes or solstices; the moonrise; moonset and the moon's eclipse; and occasionally the rising and setting of Venus. Because of the different latitudes of these stone circles, different layouts were adopted to record astronomical events.

There is another common factor. All building on European stone circles ceased by 1450 BC, when Thera's volcano erupted, destroying the Minoan civilisation.

28

STONEHENGE: THE MASTER WORK

In the previous chapter I mentioned Sir Isaac Newton's thesis that new ideas can only develop following an initial leap of faith. Newton also said that nothing made his head ache so much as accounting for the motions of the moon.

Stonehenge is the perfect place, on a clear midsummer night, to develop an obsession not just with the moon but with the amazing sweep of the whole night sky. Stand on any nearby hilltop, the breeze in your face, and let your gaze sweep the landscape. Drink in the ceremony and drama of the henge itself; its solidity, its permanence. Look harder and if it's still light enough you can see grand, ancient ceremonial avenues, *cursi* and hundreds of burial barrows. Stonehenge is magnificent, sacred and sublime, the bare bones of Britain's prehistory.

People have been worshipping here, it's thought, since 7200 BC. Stonehenge itself – the name probably comes from the Saxon *stân* meaning 'stone', and *hencg* meaning 'hinge', or 'hanging' – was built in three main phases. The first probably dates from 3000 BC to 2920 BC, when people dug out a roughly circular enclosure about 100 metres (330 feet) in diameter, surrounded by a ditch or inner bank, which was built of the earth from the ditch. There were two entry points to this initial enclosure, from the northeast and the south-west.

The ditch was crudely made. In the 1920s Colonel William Hawley, one of the pioneering archaeologists at Stonehenge, likened it to 'a string of badly-made sausages'. It was dug using picks made from the antlers of red deer and spades made from their broad shoulder blades – the same implements that had been used to dig the Great Orme Mine. The site has been carbon dated from these bones. It seems that the original purpose of this first site was ceremonial – as a gathering place to celebrate the arrival of spring, and possibly as a cemetery for the dead.

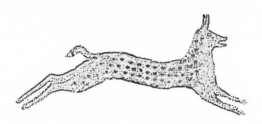

The next phases changed Stonehenge's character completely, because the builders used stones. Vast stones. For reasons we don't fully understand, around 2500 BC these huge lumps of rock were brought to the site, probably from the Marlborough Downs 23 miles (37 kilometres) away. Known as sarsens, they were cut from dense, durable silicified sandstone. Each stone erected in the outer ring was about 4.1 metres (13 feet) high and weighed around 25-tons. The large stones in the inner ring – ten uprights and five lintels – weigh up to 50 tons each and are linked to each other using complex jointing techniques. Stonehenge's builders set up the massive sarsen uprights to tolerances of just a few centimetres. It makes you marvel: how was all of this done?

An even greater mystery is how the smaller stones, collectively known as blue stones, were brought here. Although there is a theory that glacial action brought them near to the site, this seems highly unlikely. The 2- and 4-ton stones seem to have come from the Preseli Hills of south Wales – a full 150 miles (240 kilometres) to the west of Stonehenge (see map). Intriguingly, as the crow flies Preseli is about 100 miles (160 kilometres) south of the Great Orme mine, which was in full swing by this time. One new theory, supported by radiocarbon dating, has it that the stone construction

The Hagia Triada arcophagus, Crete.

Clockwise from top left: Sculpted stone sunflower juxtaposed with live sunflower, Halebid, Karnataka, India; Stone carving at Pattadakal temple, India, shows a parrot perched on a sunflower; Stone carving of a pineapple in a cave temple in Udaiguri, India; Wall sculpture from Hoysala Dynasty Halebid temple at Somnathpur, India, showing maize ears.

Clear similarities in the custom of bull leaping, which transcends cultures and countries. *Top:* Engraving by Goya, 'The Agility and Audacity of Juanito Apiñani in the Ring at Madrid'. *Middle:* Minoan bull leaper, British Museum. *Bottom:* Minoan bull leaper, Heraklion Museum.

The Dover Boat at the Dover Museum and Bronze Age Boat Gallery.

The mysterious Nebra disc, found in Germany, 1999.

England's most celebrated stone circle, Stonehenge, Wiltshire.

An amber necklace from the Upton Lovell Bronze Age hoard,
Wiltshire Heritage Museum, Devizes.

A selection of artefacts found near Stonehenge and in the surrounding area,
courtesy of the Wiltshire Heritage Museum, Devizes.

Comparisons of copper tools and implements found in the Mediterranean and at the Great Lakes, USA:

Coiled snake effigy (Uluburun wreck and Great Lakes).

Animal weights at the British Museum.

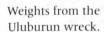

Weights from the Uluburun wreck.

Weights found in the Great Lakes area of the USA.

Conical points (Uluburun wreck and Great Lakes).

Triangulate spear heads (Uluburun wreck and Great Lakes).

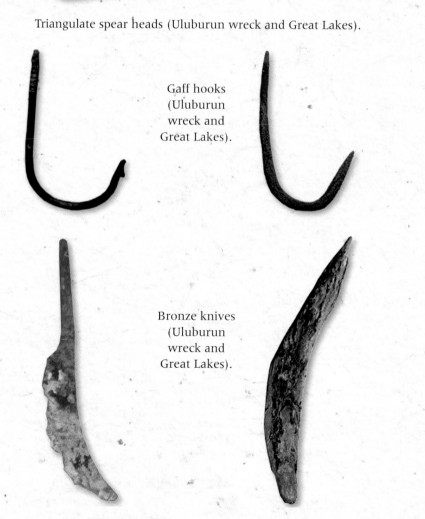

Gaff hooks
(Uluburun
wreck and
Great Lakes).

Bronze knives
(Uluburun
wreck and
Great Lakes).

Skeleton known as the 'Amesbury Archer', now in Salisbury and South Wiltshire Museum.

The Antikythera device, National Archaeological Museum of Athens, Greece.

The Isopata ring, Heraklion Museum, Crete.

at the henge began between 2400 and 2200 BC. The change of date intrigues me.

This new survey by Professor Tim Darvill and Geoffrey Wainwright contends that the smaller stones were arranged in two concentric circles within an older wooden enclosure. The big sarsen stones were now carefully trimmed with bronze axes, adzes and saws to produce sharply defined rectangular blocks with mortise and tenon and tongue and groove joints to lock the stones together. The final arrangement of the stones was almost certainly completed sometime between 2280 and 1930 BC.[3]

The scenes here at the summer solstices of the 21st century AD – with hippies banging drums and the Wiltshire police force's new aerial drone sweeping back and forth, lights flashing, filming the crowds from a hundred or so metres (a few hundred feet) in the air – may seem worlds away from what would have happened in the era before Christ. Yet perhaps it is not all that different in spirit. The crowds all come for the consuming drama of one moment – when the sun rises from behind the magnificent Heel Stone. Somehow, for a magical few minutes, it feels as if the sun has stopped dead in its tracks and hangs suspended in time.

The structure that was completed sometime between 2280 and 1930 BC was sophisticated. For example, the horizontal stones that were locked in place on top of the upright ones were not perfectly rectangular. The stones are gently and deliberately curved on both inner and outer faces. As a result the stones appear as a perfect ring, suspended high above Salisbury Plain. This effect, playing with the perspective supplied by the human eye, may well be a technique imported from the Mediterranean. The Minoans certainly used it and they handed it down to the Greeks, whose word for it is *entasis*.

From a navigator's point of view, the circle of horizontal stones provided a perfect artificial horizon. That would have enabled astronomers to note precisely when the moon rose, or which was the first star to rise in the east after sunset. It would also allow them to time both solar and lunar eclipses accurately and determine the exact times of sunrise and sunset.

The Minoans had arrived in Britain by 2300 BC to collect tin and their technological influence can be seen in bronze implements dated 2200–2000 BC, when the percentage of tin content in those implements leaps to 11 per cent.[4] At some stage (it is hard to prove definitively when) five huge stones were repositioned at the centre of Stonehenge, in a very similar way to the central stones at Nabta. So we have builders who could transport massive stones hundreds of miles, then prepare them with bronze tools for astronomical purposes, using a similar layout to the one found at Nabta.

Could the Minoans have been behind that change? It's their appetite for luxury that gives them away. I am here at Stonehenge armed with the knowledge that oriental cowries and jewellery, including blue-glazed and glass beads very similar to those produced on ancient Crete, have been excavated from a number of Bronze Age graves at Stonehenge. They are, according to Professor L. Augustine Waddell, 'of the identical kind common in Ancient Egypt within the restricted period of between about 1450 BC to 1250 BC'. The Uluburun wreck's cargo included almost identical examples of blue glass, cowries and amber.

Was this idea too incredible? Could I realistically expect to challenge the official version of history, especially when it came to one of Britain's most loved, most written about, most iconic monuments? The fact is that the more one peers into the mists of Stonehenge, the more the evidence of Minoan influence appears from the murk.

The most exciting initial discovery? During a casual Internet search, I chanced across an old Australian blog. The blogger gave an account of his tour of Stonehenge in the 1950s. You could still walk among the stones at that time, and he described a particular beam of light that had allowed him to see a carving of a double-headed axe ... etched along the stones. I immediately thought back to Almendres and another tourist's hastily shot photograph showing what looked to me like a carving of a *labrys*, a Minoan double axe, its shape eroded over the many centuries that have passed. I'd scarcely paid any attention to it. Then it clicked. I remembered Knossos, and what archaeologists call the 'mason's marks' engraved on its stones – again in the form of Minoan double-headed axes.

Here is the remarkable truth. As soon as I realised that there was an exciting lead to track down, I began looking for answers. Within just hours, thanks to the wonders of the Web, I'd discovered that there is not just one such 'axe' carving on the megaliths of Stonehenge: there are many. They were created nearly 4,000 years ago. Some have interpreted these weathered, axe-shaped marks – a 'stalk', as it were, supporting a horizontal line – as mushroom shapes. However when archaeologist Richard J. C. Atkinson drew them in 1953, he realised that one of these centuries-old carvings was probably a dagger, the other a double-headed axe. What is also remarkable is the fact that these carvings were rediscovered in the last century – and yet fifty years on, they had still not been studied and were scarcely even recorded.

With the British weather fast eroding their outlines the marks are rarely seen today, if only because no one is allowed close enough to the monument to photograph them – except during the chaotic celebrations that annually mark the summer solstice. What with the impact of wind and rain, time was, in a sense, running out.

In 2002 Wessex Archaeology contacted Archaeoptics Ltd of Glasgow, which is a specialist company at the leading edge of applying pioneering new techniques to archaeology. The year before, the company had laser-scanned the timbers of Seahenge, the intriguing Bronze Age circle of waterlogged wooden posts exposed by the sea on a remote beach in Norfolk. The results had been high-resolution, digital 3-dimensional (3D) models – most helpful for analysis.

Wessex Archaeology decided to investigate the potential of laser-scanning the Stonehenge carvings. Specialists Alistair Carty and Dave Vickers travelled to Wessex Archaeology's Old Sarum head-quarters near Salisbury with an impressive array of equipment, including a Minolta VIVID-900 scanner, capable of capturing mil-lions of points in 3D and taking measurements just microns apart. The team included Wessex Archaeology's Thomas Goskar, also a specialist in digital techniques.

The surfaces were photographed and scanned at a resolution of 0.5 mm, creating hundreds of thousands of individual 3D measure-ments known as a point cloud, which could then be animated into a 3D solid model.

This is Goskar's description of the images scanned on 'Stone 53', one of the famous sarsen trilithons:

> The first carving is 15 by 15.3 cm [5.9 by 6 inches], with a broad upturned blade, and a form of 'rib' a third of the way down the length. Although further analysis is needed, this shape could represent two axes, one carved over another. The second carving, 10.6 by 8.6 cm [4.2 by 3.4 inches], is very faint indeed, but seems to be a normal flanged axe, as we find elsewhere on the stone ...

> There was something poetic about the juxtaposition of the most advanced Early Bronze Age technology, with the most advanced 21st century archaeological recording methods. What was intended as an investigation into how well the carvings would be recorded by a laser scanner, turned into a major discovery.

> We must remember that while the sarsens are thought to have been erected around 2300 BC, metal axes were not in common circulation for generations after this. Whatever the carvings mean, accurate recording is vital to our understanding of the monument as a whole.[5]

EARLY RESEARCH

During the early dawn of the British archaeological profession in the 18th century, the antiquarian William Stukeley discovered what local people had probably always known. That at midsummer sunrise the first of the sun's rays shine into the centre of the rings of stone, meeting between the open arms of a horseshoe of stones. In fact, in the early 20th century Sir Norman Lockyer argued that there was a ritualistic connection between Stonehenge and sun worship. The precision of the alignment of these ancient megaliths with the sun could not be an accident.

Then Dr Gerald Hawkins, a well-known astronomer from America, burst upon the scene. Hawkins was Professor of Physics and Astronomy at Boston University in Massachusetts. In 1962 he and his assistants filmed sunrise at the summer solstice at Stonehenge. They plotted every stone and pit on the site and fed their co-ordinates into an IBM 704 computer, then the world's most powerful – computers being in their infancy. The journal *Nature* published Hawkins' first results in 1963.

Hawkins argued that the computer results proved that Stonehenge was a giant observatory for predicting eclipses of both the sun and the moon. His claims generated huge publicity. Professional archaeologists were furious. Here was an astronomer, and an American to boot, trampling over their patch and using new-fangled, unproven computer methods to uncover the secrets of 'their' beloved Stonehenge. Richard Atkinson described Hawkins' argument as 'tendentious, arrogant, slipshod and unconvincing' – to Atkinson, the builders of Stonehenge were 'howling barbarians'.

The archaeologists had overplayed their hand. It was obvious that Hawkins knew his subject – Stonehenge was his sixty-first published paper. Then again Hawkins was not, in fact, an American, but came from Suffolk. His degrees were in physics and pure maths and his PhD in radio astronomy was obtained at Manchester

University. Hawkins had forever changed the way we think of Stonehenge.

Next into the fray was Sir Fred Hoyle, the most highly respected British astronomer of the day. He examined Professor Hawkins' research and went even further. To him, Stonehenge was a model of the solar system. Hoyle selected three stones representing the sun, the moon and the moon's orbit. These stones were then rotated around the holes of the Aubrey Hole ring relative to one another.

Hoyle could show that when the three marker stones lay either close to each other or opposite each other, eclipses would take place. The eclipse would occur when the moon stone was closest to the sun stone or was precisely opposite to it on the other side of the Aubrey Hole ring. Hoyle's method is more accurate than Hawkins', because Hoyle's system could predict the actual day of lunar eclipse nineteen years into the future. Hoyle also identified many other astronomical alignments at Stonehenge. Hoyle and Hawkins had to take a leap of faith, based on the evidence before them: they did not have all the detailed proof we now have about the sophistication of Babylonian, Minoan and Egyptian astronomy, a body of knowledge that already existed during the latter stages of the building of Stonehenge. Neither did they have proof that the Bronze Age world was as elaborate and sophisticated as we are only just now beginning to understand – nor that there was long-standing contact between the civilisations of the Minoans, the Egyptians and the Babylonians.

From the Egyptian astronomers at Nabta, the Minoans could have known of the sun's daily maximum elevation and declination. From the astronomers at Giza we know that the Egyptians also knew about the earth's precession – from the apparent precession of Kochab. The Minoan mariners may have learned much more from the Mesopotamian astronomers, not least the precise times of the moon's rising – something that, if Hoyle was right, could also be measured at Stonehenge. Had they had a good clock they could have calculated accurate longitude from the moon's eclipse. From the altitude of the sun, taken each day at its maximum elevation, they could have deduced declination to determine latitude.

29

FROM THE MED TO THE MEGALITH

Supposing for a moment that the builders of Stonehenge did complete phase III of the circle around 1750 BC. If there really had been that much Minoan influence on the extraordinary monument, then logic would suggest that the travellers would have left evidence from the eastern Mediterranean: goods, perhaps, or traces of trade or even physical habitation.

To quote Professor Hawkins again:

> Archaeologists are traditionally conservative and ungiven to theorising, but the indications of a Mediterranean origin for Stonehenge [phase III] are so strong that they allow themselves to wonder if some master designer might not have come all the way from that pre-Homeric but eternally wine dark southern sea [the Mediterranean] . . .

R.J.C. Atkinson inclines seriously to this theory, making much of the evidence of dagger and axe carvings and Mediterranean artefacts in the burials of Stonehenge.[6] Atkinson's views are supported by a number of distinguished historians. For instance, according to Professor W.J. Perry:

> Megaliths [stone circles] all over the world are located in the immediate neighbourhood of ancient mine workings for tin, copper, lead and gold or in the area of the pearl and amber trade.

As Herodotus wrote: ' . . . it is nevertheless certain that both our tin and our amber are brought from the extremely remote regions in the western extremes of Europe'. [7]

Had the Minoans left any distinctive 'calling cards' on the rolling plains of Wiltshire? I set out for Stonehenge on a fair September day and enjoyed the sight of the stones as they leapt majestically into view from the much despised A303 road.

It's fair to say that until recently the wider landscape around Stonehenge, which is mainly rolling farmland, has been pretty much ignored. The long and bitter public dispute about burying the road, which roars past the stones on its way to the West Country, has dominated the Stonehenge debate to the detriment of other things. Yet it is obvious now that this special area was a vast, interconnected sacred landscape. The largest prehistoric mound in Europe, Silbury Hill stands just one mile (1.5 kilometres) to the south of the monument, flanked by the West Kennet Long Barrow. Avebury's stone circles are just 15 miles (24 kilometres) to the north. There is a lot more yet to discover.

Kings and important figures were buried in round mounds called barrows and many of them overlook the stones. Of late the archaeological pace has been increasing, with some spectacular finds of Bronze Age tombs. The so-called 'Amesbury Archer' was only discovered in 2002, when a new housing development was begun at a nearby village. The press dubbed him the 'King of Stonehenge', because the goods found buried with him were so rich in quality.

Twenty-nine groups of exhibits came from the barrow group of Winterbourne Stoke. The earliest is a Neolithic long barrow. More than a thousand years later, it was followed by a line of large bowl and bell barrows. The barrows are a mile southwest of Stonehenge, lying in a NE–SW line. Access is easy; there is a lay-by on the A303.

Almost all of the barrows were excavated in the early 19th century, though some were investigated in the 1960s. The most spectacular finds were the remains of two wooden coffins, many decorated pottery vessels and spearheads and daggers of bronze.

I arrived after a brisk five-minute walk through the woods to see pairs of disc, bell and pond barrows and nineteen bowl barrows. Each of the round barrows held one body – presumably a chief living at Stonehenge. The excavations yielded daggers, knives, awls, tweezers, cups, amber, dress pins, faience beads, food containers and urns.

For me, standing beside this line of barrows was a moment of *déjà vu*. I'd seen burial tombs just like them near the palace of Phaestos in southern Crete, where our adventure began. They lie on the foothills of the Cretan mountains, inland from Phaestos, on the Mesara plain. The tombs of Mesara are constructed of local stone bound together with mud, the roof supported by a type of corbel construction. They are the same height and circumference as the barrows at Winterbourne Stoke.

This could of course be coincidental. To be more certain of a cultural link I needed to go and see the actual grave goods that had been found in the barrows. They included armour, weapons, farming equipment, woodworking tools, jewellery and household and domestic utensils: many are on display in local museums. The two key museums are the Wiltshire Heritage Museum at Devizes and the Salisbury & South Wiltshire Museum, opposite England's loveliest cathedral.

These marvellous museums have thousands of Bronze Age artefacts, especially from the hoards buried at Wilsford (see map), Upton Lovell, Winterbourne Stoke, Amesbury and Wimborne St Giles. The extremely helpful director of the Salisbury Museum, Adrian Green, and the equally helpful director of the Wiltshire Heritage Museum, David Dawson, kindly let me photograph the exhibits.

I separated the Early, Middle and Late Bronze Age artefacts buried in the barrow mounds into twenty categories. The main ones were axes; adzes; jewellery; personal hygiene; woodworking tools (chisels, hammers, etc.); farming implements; dress and couture; hunting equipment; weapons offensive and defensive; votive offerings; ceremonial items (mace, Minoan double-headed axe); games and pastimes;

trade goods (balance weights); and kitchen and cooking equipment. These twenty principal categories were then further broken down – kitchen equipment into pots and pans, cups, knives, spoons and so on.

I then placed photographs of artefacts in these twenty categories beside similar items found in the Uluburun and Gelidonya wrecks. (I established that the objects found in the wrecks were in fact Minoan in the manner described in chapter 37.) The results may be seen on our website.

The results speak for themselves (see second colour plate section). The people buried at Stonehenge in the Bronze Age used the same bronze weapons as their Minoan counterparts – tanged and riveted knives, swords, lances and arrows, the blades often with the same ornaments. The simplest explanation is that the objects were in fact Minoan.

The presence of Minoan-type artefacts at Stonehenge does not mean that both civilisations had reached the same stage of development and that these items were made by Britons. Anyone who has visited the magnificent palaces of Knossos and Phaestos becomes conscious of the huge gap between ancient British and Minoan culture. The building technology alone reinforces this point: were there any similar Bronze Age palaces in Britain? The answer is a resounding no.

Amber has been found in twenty-nine graves in the area, not least in the form of some exceptionally valuable necklaces. This jewellery is often Minoan in character. Archaeologists have had some of the amber scientifically examined – and I was not surprised to find it comes from the Baltic. Between the amber pieces in many of the Stonehenge burials are amber spacers and faience beads made of crystal or glass. Archaeologists already accept that at least some of these came from the eastern Mediterranean. Faience beads of the same shape and colour were found in the Uluburun wreck. The ceremonial mace at Stonehenge had its counterpart at Mycenae, and so on. The plot was beginning to thicken.

Seen in that light the close similarity of awls, knapping tools, bracelets, armbands, scales, knives, twisted bow drills, triangulated socket points, spades, daggers, necklaces, earrings, bangles, torcs, rings, brooches, earlobe adornments, cups, plates, lance heads, chisels, spearheads, gaff hooks, weights, pins, buttons, fasteners, cleavers, hammers, saws, bradawls, drills – thirty-two separate types of artefacts – cannot be a coincidence. The most exciting of all these ancient things? Two Minoan double-headed axes in the local museum. Just like the famous *labrys* of King Minos, found at Knossos.

Was Stonehenge a holy site for the Minoans? Perhaps even a place of pilgrimage? Had they, as part of the long-term trading agreements they held with the local Britons, begun to settle here?

30

THE LAND THAT TIME FORGOT

It was at Callanish on the lonely Isle of Lewis that I suddenly had a breakthrough. I realised that, in a manner of speaking, I'd been barking up the wrong tree. The hoard I was here to see, one of the finest Bronze Age hoards ever found in Scotland, contained Irish bronze tools and beads of Irish gold. But it was the trees, not the hoarded bronze and gold that I began thinking about. More precisely, it was the lack of them. It was then I realised that my thinking about Minoan journeys to the west had been, in effect, the wrong way round.

Today, Lewis has large areas of peatland and bog. But reading my guidebook, I discovered that until about 1500 BC the island had been warmer, more fertile and much less wet. I suddenly realised that it was deforestation as well as climate change that had denuded the land. Deforestation. The kind that happens after shipbuilding and repair; the kind that happens after large, industrial-scale smelting operations ...

At first sight, it seems highly unlikely that Minoan ships from sunny Crete would have ever ventured to the stormy, wet wilds of the Hebrides. But there were many more of these mysterious Bronze Age stashes hoarded in this general area – at Gurness on the Orkneys, for instance and at Dunagoil on Bute. Why, here, on this wild and windy series of islands? What did these sites have in common? One fact was immediately obvious: they had all been built before widespread local deforestation. Was that purely happenstance?

Had this remote island had a population large enough to carry out an elaborate build, such as the Callanish stone circle? It seemed unlikely. Or had the islanders had help when creating the stone ring – in the form of either hostile invaders or very persuasive outsiders? Was there a lack of trees because large-scale smelting had taken place on these islands?

There was one stunning new piece of evidence: DNA. Examining the DNA of the island's current population, I discovered that Lewis had a high frequency of haplogroup X2; and one of the few other places that had a similar high incidence of haplogroup X2 was – Crete.[8] (DNA is described in more detail in chapter 38.)

I was seeing a repeating pattern of activity. Could this idea be true? The connections seemed to be there. Besides the Irish wares, the hoard at Adabrock on Lewis also contained the Minoans' characteristic calling cards of amber from the Baltic and glass from the Mediterranean. That was all well and good. But the inevitable question arose: what had been the Minoans' real interest in the windswept Atlantic island of Lewis?

The true scale of the Minoans' trading ambitions was beginning to dawn on me. I was standing at the epicentre of what had been a network: I was unveiling a true trading empire, with bases and ports set up along the entire route. This empire hadn't simply straddled the eastern Mediterranean. The people of Crete had been the East India Company of their day, a maritime enterprise whose size and ambition took your breath away. Where those trading colonies or bases would be was defined by a number of factors – the frenetic trade in bronze, and the desperate need for exploration, to keep that river of bronze going.

I had been hoping that later at night I would watch the moon skim the stones, setting them aglow. Although I was unlikely to see the phenomenon that Diodorus Siculus had described, 'the god' of the moon 'visiting the island'. Sadly, I was here at the wrong point in the lunar cycle: I would need to wait until 2034 to observe that almost magical event.

It kept niggling at me as I walked. Why here? Why would the Minoans come all this way, up to the northern tip of the Outer

Hebrides? It was then that it struck me. What if they hadn't been coming to Lewis, like tourists? What if they'd actually been *en route* to somewhere else? The idea struck me like a blow to the head. What had I been thinking all this time? I'd been looking at things the wrong way round. When they reached Callanish, our ancient travellers had been *returning home* from a much longer journey – one they had made on the wings of the Gulf Stream. Forget Christopher Columbus. It was the Minoans who had first discovered America.

IIIIIIIIIIIIIIIIIII

THE JOURNEY TO AMERICA AND BACK

If Callanish had been the landfall for the Minoan traders returning from America, where did they start from? If the sailors overwintered in Spain, ships intending to make an Atlantic crossing in the favourable season of May–August could begin their voyage from an Atlantic port such as Cádiz or San Lúcar de Barrameda, where Bronze Age ports have in fact been found. Put another way, ships from Thera could winter somewhere like San Lúcar before starting an Atlantic crossing the following year. Overwintering in San Lúcar they could have repaired any broken steering oars and sails and careened the ship to remove barnacles from the hull. In May, past the hurricane season, they could use sail, rather than oar power, and the massive force of the ocean currents to take them to new-found lands.

Then, loaded with copper and following the huge circle of the Gulf Stream, they could speed their way *back* to Britain from America. On the return journey, reaching landfall at Lewis would have meant being halfway for the Minoans. The Gulf Stream is a massive force to be reckoned with. This colossal current of moving water is caused by the earth's rotation. The great stream flows clockwise all the year round in the North Atlantic, bringing the warmth and fertility of orange-growing Florida northwards with it. It's part of a whole network of currents; the Spanish used their sublime force and power to reach the Caribbean, in their treasure-hungry galleons.

In Britain we have the Gulf Stream to thank for our relatively warm weather. When the American statesman Benjamin Franklin analysed its properties in the 1760s, he managed to knock weeks off the standard sailing time between America and Britain by refining the routes ships would take, employing the forces of nature rather than struggling against them. When the Minoans discovered the Gulf Stream, they must have realised that it gave them an express train ride, for free.

In 1970, as captain of the submarine HMS *Rorqual*, returning from America to Scotland, I requested permission to vary my sailing orders so that I could pass through the Denmark Strait and then through the Faroes Gap and on to northeast Scotland. I wanted to travel with the Gulf Stream and see when it petered out.

A submarine is the ideal vehicle with which to measure the power of this mighty river of warm water as it carves its way across the Atlantic. A submerged submarine needs to equal the weight of the water it displaces. In hot water, the submarine must lighten its weight. If the water then gets colder, the submarine needs to flood in seawater. So when the warm Gulf Stream water disappears – if, say, the submarine is below it at 152 metres (500 feet) – this will become apparent from the submarine's weight. We discovered that in terms of volume the Iceland–Faroe flow is the strongest of the three current branches flowing from the Atlantic Ocean into the Nordic seas, across the Greenland–Scotland Ridge.

By the time they got to Lewis, on the extreme western edge of Europe, laden with copper from America but desperate for food and water, the Minoan crews would have needed to rest up, repair their ships and restock.

The islands would have become a pivotal point in a powerful Minoan trading empire that spanned the entire Atlantic. From here, these enterprising explorers would have been able to launch further lucrative trading missions to Denmark, Greenland and beyond. These were consummate businessmen: they would maximise their reach to maximise their profit.

As Plato had said: 'This power came forth out of the ocean.'

It was as I had suspected: the Minoans' tremendous seafaring capability had given them control of a vast trading empire; an empire far larger than simply the Mediterranean. It was one that took full advantage of the vast mineral wealth of the west, of America, 'an island larger than Libya and Asia together'.

‖‖‖‖‖‖‖‖‖‖‖‖‖‖‖

The Minoans probably did not set out to colonise the Outer Hebrides or the Orkneys, but arrived there while heading for home, ships filled with pure American copper (see chapter 33). With that in mind we can look at what they left behind them: starting with voles.

There is clear evidence of the human introduction of three types of vole in northern Europe: the field vole to the Outer Hebrides, the bank vole to Ireland and the sibling vole to Svalbard. None of these voles are found on the mainland of the British Isles.

Which group of humans brought these voles, and when? Radio carbon dates of $3,590 \pm 80$ (i.e. c.1500 BC) and 4800 ± 120 (i.e. c.2700 BC) are from two bone samples of voles excavated in the Orkneys.[9] DNA comparisons made by York University show the closest match to the Orkney vole are those of southern France and Spain. So it seems seafarers came from southern France or Spain to the Orkneys between 2700 BC and 1500 BC. Rodent stowaways in Minoan ships would have been commonplace – the Uluburun wreck had a Syrian mouse! An alternative explanation is that the voles reached the Orkneys first, then southern France and Spain.

The common vole is the only vole on Orkney. It inhabits eight Orkney Islands: Burray, Eday, Mainland, Rousay, Sanday, South Ronaldsay, Stronsay and Westray. The vole cannot swim, so it must have been brought by humans. It is most likely that they were in the hay or straw which seafarers took along for animals. This vole is not found on the adjacent Shetland Islands, nor on mainland Scotland or in England. It is at least arguable that the voles link the Orkneys with southern France and Spain.

CONCLUSION

Because of their physical location at the end of the Gulf Stream, the Outer Hebrides and the Orkneys became trading hubs for Minoan ships bringing copper from the Great Lakes to Europe. Perhaps burials in the Orkneys would tell us more?

A BBC television programme contained this report by David Keys:

> According to sensational archaeological discoveries currently being made in Scotland, Bronze Age Britons were practising the art of mummification at the same time as 'mummy culture' was in full swing in Pharaonic Egypt. It appears that ancient Britons invented this skill for themselves ... [10]

A team of archaeologists, led by Dr Mike Parker Pearson of Sheffield University, had made an astounding discovery on the Hebridean island of South Uist. Two mummified bodies were buried under the floor of a prehistoric house in an area called Cladh Hallan. The house was part of a unique Bronze Age complex. The report said that the complex was made up of seven houses arranged as a terrace, and 'is as mysterious as the preserved corpses that were buried there'. The report went on:

> To the astonishment of the archaeologists, they saw that one individual, a male, had died in around 1600 BC – but had been buried a full six centuries later, in around 1000 BC. What is more, a second individual (a female) had died in around 1300 BC – and had to wait 300 years before being interred ...

The report speculated that the bodies could have been members of some ritual élite, potentially priests or shamans. It is just as possible that they were new arrivals or settlers. I suggest the people were Minoan leaders. The Minoans knew of mummification from the Egyptians, notably from their extended stay at Tell el Dab'a. It's worth mentioning that when he was excavating at Mycenae,

Heinrich Schliemann noted that one of the bodies he uncovered had been mummified.

The extraordinary thing was that the more I looked, the more there was to find. We will seek to have the DNA of the South Uist mummies tested to see whether the haplogroup X2 is in their genes. The intriguing fact about haplogroup X2 is that it has been found not just in the Orkneys and not just where the Minoans originated (see chapter 38) but in the Americas of the Great Lakes. Nowhere else is this haplogroup so prevalent and so highly marked in local populations of today. As Professor Theodore Schurr states, 'A genetic marker appropriately called Lineage X suggests a definite – if ancient – link between Eurasians and Native Americans.'[11] So the scientific evidence backs up the high incidence of shared DNA between Orcadians and Cretans: both have a very high incidence of haplogroup X2 (7.2 per cent) in their genes.

31

THE BRONZE BOY

Now, I realised, the Minoans ran a vast Bronze Age 'Common Market', stretching from the Orkneys right through to India. They took English copper and tin all the way to ancient Egypt to make the Saqqara Pyramid's bronze saws. However, as time went on, the copper and tin became scarce. But if the Minoans fuelled the entire global trade in the raw materials and finished products of the Bronze Age, then where else could it have been sourced? I already had an idea. But before we leave Britain's shores to brave the Atlantic, there is one last thing I would like to do: like a million or more others, I am going to return to Stonehenge.

Only last year, in 2010, came a rather stunning discovery. Scientists from the British Geological Survey had conducted tests on the body of a fifteen-year old boy – a body found in 2006 at Amesbury, quite near to the henge. The boy had been placed in a simple grave just a mile from the stone circle, an amber necklace of huge wealth – with ninety beads – by his side. The tests prove that he died of infection, rather than from any violence. More to the point, a broad range of scientific studies prove the teenager was from the Mediterranean.

The discovery adds further meat to my overall thesis. The Mediterranean boy was buried here in around 1550 BC, a date that has significance. He could have been here as an apprentice, to pick up and capitalise on the Minoans' lucrative system of established trade routes. Alternatively, he could have been a Minoan fleeing the Mycenaeans' invading armies.

I'm convinced that Stonehenge was a prehistoric international landmark. In the words of project archaeologist Andrew Fitzpatrick:

We think that the wealthiest people may have made these long-distance journeys to source rare and exotic materials, like amber. By doing these journeys, they probably also acquired great kudos.[12]

The most spectacular spiritual sanctuary in the world, Stonehenge would have been famous beyond measure, because reading the heavens and learning how to travel the world was tantamount to reading the will of the gods. It was a religious rite.

Other people who had visited Stonehenge from afar include individuals found in a collective Bronze Age grave, the 'Boscombe Bowmen', who were almost certainly from Wales.

Crowds came here initially for the religious ceremony and observation of the stars. But where better for the bringers of bronze to conduct trade than at a place where thousands gathered in prayer? Bronze Age wealth transformed the entire plain into a major centre of commerce and exchange.

At that other mystical stone circle, Callanish, I had already made the connection between the Minoans and copper-prospecting in the Americas. Now here, at Stonehenge, I was beginning to think there could be much more evidence of transatlantic trade. This grave and the body of this teenager weren't found until the 21st century: how much more, then, is left to find out?

In May 2002, Wessex Archaeology was doing a routine excavation in an area that was due to become a housing scheme at Amesbury, a few miles from Stonehenge. It was simply standard procedure in an area like this to look, before the whole thing became covered in concrete. If they found anything at all, the Wessex team were expecting to come across Roman remains. But it wasn't long before the archaeologists found something that turned their whole view of this place on its head. What they'd uncovered was a grave. No ordinary burial, this outwardly unassuming site was filled with pottery dating back at least 2,500 years before the Romans arrived in Britain.

By mid-afternoon the team had literally struck gold: gold

jewellery. A bank holiday was due that weekend and there was a danger that the grave would be robbed or disturbed if it was left unattended. The archaeologists knew they couldn't leave this find, so the diggers worked on through the night, with just their car headlamps for lighting. By dawn they had unearthed the entire skeleton of a man. His was the richest Bronze Age grave ever found in Britain.

The body of the man, also known as the 'Amesbury Archer', has since been dated to around 2300 BC, the early Bronze Age, almost a thousand years earlier than the burial of the Mediterranean 'Boy with the Amber Necklace'. This was around the time that the first metals were brought into Britain. His mourners had laid the 'Archer' on his left-hand side, with his face to the north. Buried alongside him were the weapons of a hunter, including three copper daggers. Bound to his wrist was a slate guard, to protect him from the snap recoil of a bowstring.

Clearly, this was a man of high status. Only the non-organic objects he was buried with survive, so we don't know what he was wearing, but with him were two beautifully worked earrings in gold and two golden hair ornaments.

The archaeologists' dating places him here at the exact same time that the massive stones were being erected at Stonehenge. And as he was buried less than 3 miles (5 kilometres) away, there is speculation that this man of special status had had a hand in the planning of the monument.

Later, another grave was found close by. Curled up in a foetal position and lying against the pockmarked chalk of his grave, this younger man was almost certainly related to the first. Both skeletons had the same highly unusual bone structure in their feet – the heel bone had a joint with one of the upper tarsal bones in the foot. It is possible that they were father and son.

When he died the 'King' was thirty-five to forty-five years old. He was buried with objects useful to him in the next world, including arrowheads and copper knives. His cushion stone, for working metal, lay right next to him. He may have been one of the first people in Britain to have been able to work gold: hence the richness of his grave.

His teeth were examined by oxygen isotope analysis, which can help identify where a person lived when he or she was young. Stronger than bone, tooth enamel is the hardest and the most mineralised substance in the human body – one of the reasons why human teeth can survive for centuries after a person has died. It envelops the teeth in a protective layer that shields the underlying dentin from decay. The enamel grows quickly until puberty and holds a chemical record of a child's environment, even stretching to the climate and local geology.

The enamel's chemical components are mainly calcium, phosphorus and oxygen, with trace amounts of strontium and lead. Of these it is the isotopes of oxygen and strontium that are the strongest indications of the climate where a person grew up. The ratio of heavy to light oxygen isotopes depends on the water you drink when you are young. Drinking water in a warm climate results in more heavy isotopes, while cold water produces a lighter chemical signature.

An analysis of the oxygen isotopes within the dental enamel of the two skeletons showed that the older man came from a colder climate than was then found in Britain. The wisdom teeth of the younger man, who was between twenty-five and thirty years old when he died, revealed that he spent his youth in southern England, but then moved to the Midlands or northeast Scotland in his late teens. Because the 'King' came from a colder climate, archaeologists think he may have come from the Alps, or possibly northern Germany. I would argue that he could just as well have come from Lake Superior. Assuming he was the 'King's' son, the boy could have been left in England rather than brave the Atlantic as a child, but might have been taken to the Minoan trading posts in the Hebrides or the Orkneys as a teenager. DNA tests should determine more about the ancestry of the two men.

A famous skeleton has been found on Lake Superior in the United States, which is approximately the same age (c.2300 BC) as the 'King' of Stonehenge. The 'Rock Lake' skeleton was buried with a copper axe similar to those found at Stonehenge. We are hoping to have his DNA compared with that of the 'King of Stonehenge', to see if both are those of Minoans with the rare haplogroup X2. A number of locally found skeletons – now in Milwaukee Public Museum – have peculiar bone deformities to their feet.

|||||||||||||||||||||

There were other, independent reasons I thought I should look westwards to the Americas. I knew from enthusiastic letters and emails sent to my website that the Americas had significant amounts of copper ore. Bronze Age tools found at the copper mines of Lake Superior are remarkably similar to contemporary artefacts discovered in Britain. What's more, experts say that many of the copper artefacts found in ancient American mounds were actually produced by molten casting, a technique that was developed in the Mediterranean and was otherwise unknown in America at that time.

Tests by both the US National Bureau of Standards and the New York Testing Laboratory confirm that many artefacts found in American mounds were made using Old World casting technology. Dr Gunnar Thompson is sure that this is clear evidence of overseas contact:

Recent assays record that some of the copper artefacts found in North American burial mounds were made from zinc–copper alloys used in the Mediterranean. Ancient metal crafters added zinc to harden copper into a bronze alloy. The shapes of the copper tools found in American archaeological sites are identical to those of the ancient Mediterranean – including chisels, dagger blades, wedges, hoes, scythes, axes and spear points. These tools often have specific modifications including the use of rivets, spines and sockets. All of which were characteristic of Mediterranean tools. The fact that most of the tools were cast from molten metal implies that foreign craftsmen participated in their manufacture.[13]

If it is true that indigenous American peoples did not cast copper or make bronze, then the masses of specialist mining tools found at Lake Superior must have been made by foreigners. The foreigners can only have come by sea. If these seagoing people who sailed to America were not the Minoans, then who were they?

Notes to Book IV

1. Toby Wilkinson, *The Rise and Fall of Ancient Egypt*, Bloomsbury, 2010
2. *Complete Temples*: Wilkins on RH, 2000
3. BBC *Timewatch*, Professors Tim Darvill of the University of Bournemouth and Geoffrey Wainwright, President of the Society of Antiquaries
4. Needham, S. L. et al. 'Developments in the early Bronze Age Metallurgy of Southern Britain', *World Archaeology*, vol. 20, no.23
5. Thomas Goskar, *British Archaeology* 73
6. R. J. C. Atkinson, *Stonehenge*, Pelican, 1960
7. Herodotus 3, 115, trans. Basil Gildersleeve, in *Syntax of Classical Greece*
8. *The American Journal of Human Genetics* (AM J Hum Genet) 2003, November, 73 (5) 1178–1190–X2 of Orkney inhabitants is 7.2 (research of Helgason et al. 2001) – the second highest after Druze
9. Hedges, R. E. M., Housley, R. A., Law, I. A., Perry, C. and Gowlett, J. A. J., 'Radiocarbon Dates from the Oxford AMS System: Archaeometry Datelist 6', *Archaeometry* 29 (2), 1987, 289–306
10. *The Mummies of Cladh Hallan* BBC, 18 March 2003
11. T. G. Schurr, 'Mitochondrial DNA and the Peopling of the New World', *American Scientist*, 18 (2000)
12. Andrew Fitzpatrick, *National Geographic Magazine*, 13 October 2010
13. Dr Gunnar Thompson, *American Discovery*, Misty Isles Press, Seattle, 999

BOOK V

IIIIIIIIIIIIIIIIIIIIIIIIIIIIIIII

THE REACHES
OF EMPIRE

32

THE SEEKERS SET SAIL

It may have taken thirty days or more to reach landfall. The passing days would have turned into weeks and the crew would be getting anxious. Just how anxious is illustrated by the first nail-biting voyage made across the Atlantic by the explorer Christopher Columbus. When he supposedly 'discovered' America in 1492, a continent which he thought was Asia, Columbus kept two logs. One showed the true distance the ship had sailed on a given day. The other one he faked, to make the distance they'd travelled look greater. As the days turned into weeks, the falsified log was the only one he dared show his increasingly terrified crew.

Could Minoan ships have dealt with the rigours of an Atlantic crossing? Well, with very similar rigging the Vikings managed the same feat. Minoan sailing gear was planned for the best efficiency. Columbus' ship, the *Santa Maria* – also known as the *'Marigalante'*, or 'gallant Maria' – was fitted out with two square sails on the foremast and mainmast. She had a single lateen (a triangular, Roman-style sail) at the stern. The log details that one of the smaller ships accompanying the *Santa Maria*, the *Niña*, left Spain with lateen sails on all masts. By the time they got to the Canary Islands, Columbus had ordered her to be refitted with square sails, like the *Santa Maria*, to take advantage of the winds. The Thera frescoes show us quite clearly that the Minoans already sailed square-rigged, in the 2nd millennium before Christ's birth.

The central reaches of the Atlantic have always worried sailors, even in more modern times. Its sinister-looking masses of moving sargassum (floating seaweed); the fact that you can travel for weeks and still not reach land – all of this is deeply unsettling. In the days of sail, ships would also get hopelessly becalmed in the so-called 'Horse Latitudes', at which point the crew had to confront the very real risk of dying from thirst. You can imagine the sailors thinking: 'Is this, in fact, the edge of the world? Is this where our ship is finally going to fall off?' Truly, if 'there be monsters', they'd be here in the Sargasso Sea.

Assuming they didn't make the mistake of sailing into the Sargasso, helped by the current, the Minoans would loop up past the Antilles and on into the Gulf of Mexico. It is a monumental journey. As Columbus said of the currents of the Caribbean:

> When I left the Dragon's Mouth, I found the sea ran so strangely to the westward that between the hour of Mass, when I weighed anchor, and the hour of Complines, I made sixty-five leagues.

Strangely enough, when it comes to navigation, we can take an initial steer from the ancient palaces of Crete. We know the Minoans could calculate bearings with accuracy, not least because the alignment of all of the Minoan palaces is the same, to within just a few degrees: NNE–SSW from a datum line, or 8 degrees east of north.

The palace builders used a standard unit of measurement – the Minoan foot – as they did at their global observatory sites. J. Walter Graham, in a fascinating paper I'd searched out, studied in detail the Cretan palaces of Phaestos, Knossos and Malia and the small, Late Minoan palace of Gournia, and then put that measurement at between 203 and 204 millimetres (approx. 8 inches).[1] This alone implies that using their common unit, the Minoans had a way of calculating complex numbers and therefore distance.

I would venture further. The knowledge and sophistication of the Minoans was such that they were expert mathematicians. The late American professor and cryptologist Cyrus Gordon certainly believed this, often pointing out that in Linear A script, a small circle seemed to represent the number 100.[2] Only very recently, the Oxford University archaeologist Anthony Johnson proposed that the builders of Stonehenge were actually using Pythagorean geometry. This was a good 2,000 years before Pythagoras.[3]

Now we have certain proof. In 2010 Dr Minas Tsikritsis, a Greek mathematician, published a book about Minoan astronomy, in which he presented convincing new evidence that the Minoans had developed a solar calendar of 365.3 days. His certainty that they did so comes from his study of seal stones, rings and other ancient Minoan artefacts. Put together, the dating of those artefacts suggests that the Minoans had already developed their calendar by 2200 BC. This is around 1,700 years earlier than the Babylonian astronomer Nabu-rimanni is said to have invented the solar calendar. He did so at some point between the Persian (539 BC) and the Macedonian (331 BC) conquests of Babylon. If this is true, then it strongly indicates that the Minoans could calculate longitude – and they could do it independently of Babylon.

Here is how the Minoans calculated numbers. They had two basic forms to construct their digits: symbols which to us look like a straight line and a circle. A vertical straight line signified the number one [I], while a horizontal straight line [–] was ten. A circle [O] denoted one hundred, while a circle with four equally spaced projections [⟡] denoted a thousand. Ten thousand was represented by adding the symbol for ten to the unit for a thousand.

So a large number such as 14,266 would have looked like this:

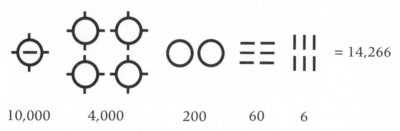

| 10,000 | 4,000 | 200 | 60 | 6 |

Thanks to Dr Tsikritsis' years of painstaking research we now know for certain that our ancient adventurers from Crete could count the number of units they reckoned they had travelled in a day – what we now call sea miles. They could record the number of days they had sailed by logging the number of sunsets since leaving Crete. So they could keep a log showing how far and in what direction they had travelled.

Now to refine our ideas about Minoan ocean navigation. They started their voyages in spring, say a month after the spring equinox (which they calculated from their stone circles). This month is described in Linear B archives by the name *'po-ro-wi-to'*. The name meant 'the month of voyages'.

There are no signposts in the open ocean. The Minoan navigators, as we've already explored, must have used a virtual reference grid – their own equivalents of latitude and possibly even longitude lines – to work their way across the world. They noted the height of the pole star and hence the latitude of their home port in Crete or Thera. This is the latitude they must reach to return.

We now know from the positioning of the stone circles at Stone-henge, Almendres and Callanish that the Minoans could determine latitude to within a mile – and they would have had to do so to avoid getting lost. Latitude is relatively easy to determine because there is a star at the extension of the North Pole billions of miles away in space. If you stood at the North Pole in 1450 BC and looked vertically above, at 90 degrees you would see the star Kochab. Because Kochab is so far away, at the equator it appears on the horizon at zero degrees; so by measuring the height of Kochab in the sky you could calculate your latitude.

Assuming this is not their very first voyage to America, our explorers know that they have to sail due west for, say, twenty days to reach the African coast at Tunis. Then they hug that coast for another fifteen days to reach the Pillars of Hercules (Straits of Gibraltar) whose latitude they know. They set off steering west by noting the position of sunrise and sunset, dividing this angle in half to find south, and checking that bearing at midday when the sun is at its highest – they then continue to steer west, at right angles to south.

At night they steer by the stars. At sunset they note the star on the western horizon nearest west and steer towards it. When that sets they pick another star on the horizon nearest to where the sun set and so on until morning, when they revert to using the sun. They duly hit the coast at Tunis, fill up with fresh water, fruit and food and sail on. When they think they are near the Pillars of Hercules they adjust their latitude to that of the Pillars by using the pole star. So far so good: they know their latitude, and they are in the Atlantic.

During this period, steering due west on the same latitude, they would have been able to check the height of the sun each midday at the same latitude – they could cross-check their latitude with Kochab. They would have noticed that the sun rose higher in the sky each midday towards summer, reaching its maximum height on midsummer's day. They would have noted that a simple correction (declination) could be applied each day to the sun's maximum measured height at midday (meridian altitude) to give latitude using this equation: Latitude = 90 ± Declination.

They could also have learned how to make these calculations at Nabta, or other stone observatories. After they had recorded the daily declination, they would have been able to use the sun to calculate latitude each day if Kochab was not visible – i.e. south of the equator. They probably ran a sweepstake; a friendly bet on how far they would travel each day.

When in the Atlantic, they turn southwest for the Canaries, where the 'conveyor belt' of the elliptical Atlantic current takes over, carrying them first southwestwards to the Cape Verde Islands, then westwards to the Caribbean. It is worth noting that cotton with chromosomes unique to North America has been found in the Cape Verde Islands. The Minoans record the time the voyage has taken so far. After resting and provisioning, the current carries them on their way northwest through the Caribbean into the Gulf of Mexico. All the time, the pole star is becoming higher in the sky. As soon as it reaches the latitude of the south tip of Florida they know they must head due west to reach the Mississippi.

They then use 'signposts' set up in earlier voyages to guide them to the rich ores of Lake Superior. They steer north up the Mississippi or west up the St Lawrence River, using these dolmen stone sign-posts.

On the return journey they again use the 'conveyor belt'. This time it takes them across the Atlantic to the Outer Hebrides (Lewis) and the Orkney Islands, where their compatriots have erected 'observatories' of stone circles, as at Stonehenge. They then follow what are now known as the English, French and Spanish coasts, with the current. Now the pole star is sinking lower in the heavens. When it reaches the latitude of the Pillars of Hercules, they know they must head due east, which they do. On entering the Mediterranean they retrace their journey to Crete, reversing their outward bound passage until they reach their joyous point of return.

Using latitude only would not have told them how far – in other words, for how many days – they needed to travel. For this they needed to be able to calculate longitude. As Charles II and the first British astronomer royal, the Revd John Flamsteed, knew, navigation is an art as much as it is a science. That art was so elusive that it became an international obsession in 18th-century Europe. In England, the struggle to perfect navigation became a long-standing joke: to be 'discovering the longitude' meant basically attempting the impossible.[4]

MINOAN CALCULATION OF LONGITUDE

Any heavenly event such as (i) an eclipse of the sun or the moon (ii) the rising and setting times of the planets (iii) the times when planets pass in front of the stars, sun or moon or (iv) the stars' rising and setting times, can be used to determine longitude, provided the observers have accurate star tables and an accurate clock. This method is described in my book *1421* at Appendix 2 pages 598–607 and in *1434* at pages 24–38.

The first requirement is for an accurate clock. Professors J. Fermor, J.M. Steele and F.R. Stephenson have summarised the inaccuracy of water clocks used by the Babylonians.[5, 6, 7] There is no way they could in practice have used them to determine longitude. However Professor Steele, in a review of N.M. Swerdlow's book, *The Babylonian Theory of the Planets*,[8] writes:

> We know that Babylonian astronomers were capable of measuring longitudes if they wished; the existence of a fragmenting star catalogue proves this. Furthermore he [Swerdlow] notes that the preserved diaries do not contain as many reports of the distance of a rising or setting planet to a normal star (from which the longitude could be obtained using something like the star catalogue mentioned above) as one would need to derive the planetary parameters. However this does not necessarily imply that such measurements were not available, or could not have been made by the astronomers who formulated the planetary theories.[9]

So how are these apparently contradictory positions reconciled? The answer I think is to do away with the water clock and rely instead on star tables, which show the rising of stars on the eastern horizon at sunset each evening for four years. At this point the cycle would repeat. In short the navigators would use the slip between sidereal and solar time, as explained in *1434*.

For example on, say, day sixty-eight the star tables published in Babylon state that Aldebaran rose simultaneously with the top tip of the sun disappearing below the western horizon. Out in the Atlantic on day sixty-eight, a second observer notes Betelgeuse not Aldebaran rose at sunset. The angular difference between Aldebaran and Betelgeuse was six hours, one quarter of twenty-four hours. Thus the Atlantic observer would know that his longitude was 90 degrees west, one quarter of 360 degrees. This eliminates the need for a clock. However, it only works if the observers are on the same latitude and if the Minoans had copies of Babylonian star tables – or had produced their own device, capable of both measuring geometrical angles and operating as a calendar.

The fact is that such a device does exist. It was discovered in a shipwreck at Antikythera in 1900; then it was locked away – and simply forgotten about.

THE ANTIKYTHERA MECHANISM

Antikythera is a tiny island just a few miles northwest of Crete. This is how the magazine *Nature* put it:

Two thousand years ago a Greek mechanic set out to build a machine that would model the workings of the known universe. The result was a complex clockwork mechanism that displayed the motions of the sun, moon and planets on precisely marked dials. By turning a handle the creator could watch his tiny celestial bodies trace their undulating paths through the sky . . .

. . . Since a reconstruction of the device hit the headlines in 2006, it has revolutionised ideas about the technology of the ancient world and has captured the public imagination as the apparent pinnacle of Greek scientific achievement.

Now, however, scientists delving into the astronomical theories encoded in this quintessentially Greek device have concluded that they are not Greek at all, but Babylonian – an empire predating this [ancient Greek] era by centuries.[10]

The importance of the Antikythera device is that it could provide planetary information – not least the position of the planets at sunset. Provided the observer at sea had the same set of tables as the observer in Babylon, the angular distance between the planets at sunset would give the difference in longitude. The Antikythera device could in fact be used as a longitude calculator – a vivid example of the brilliance of early astronomers. You can view the device in the second colour plate section.

Hoyle believed that a highly sophisticated mathematical and astronomical civilisation was behind the creation of Stonehenge. He said:

It is not until we come to Hipparchus and Ptolemy that anything of comparable stature can be found in the ancient world, and not until we move forward to Copernicus in the modern world. To paraphrase Brahms in his reference to Beethoven, we hear the tramp of the giant behind.

Stan Lusby has actually tracked those giants' footsteps. A sea surveyor and a specialist in ancient navigational techniques, Lusby used a computer programme, as he put it, to 'navigate its way through myth'. He took Homer's description of Odysseus returning home guided by the stars literally, to see whether it would have been possible to cross the Atlantic:

> The late-setting Boötes and The Bear, which we also call the Wain, which ever circles where it is and watches Orion, and alone has no part in the baths of the ocean. For this star Calypso, the beautiful goddess, had bidden him to keep on his left hand as he sailed over the sea. For seventeen days then he sailed over the sea and on the eighteenth appeared the shadowy mountains of the land of the Phaeacians.[11]

Lusby set up his 'Skymap' computer programme so that the night sky would appear as it would have done on 22 November 1350 BC, at latitude 23 degrees north and longitude 22 degrees 50 minutes west. The night sky would have been very similar to the period during which I believe the Minoans were exploring the Atlantic. The position Lusby chose to study, the point between the Canary and the Cape Verde Islands during the mid 14th century BC, is the very same course the Minoans would have steered *en route* to America. The 'Skymap' shows the perfect symmetry of the night sky at that date, with Libra and Aries on opposing horizons. Lusby argues that the ancients used 'star maps' to achieve certain latitudes then steered along that latitude – for example when voyaging south down the coast of western Europe the explorers would arrive at a latitude where Aldebaran could be seen to be vertically above Alnilam in Orion's Belt. Then they would have time to turn west into the ocean to pick up the 'conveyor belt' which would carry them to North America.

To quote part of Stan Lusby's paper, 'Odysseus, James Cook of the Atlantic':

The landfalls detected are too numerous to be confined to chance and they reveal the existence of a chart-in-the-sky for the North Atlantic that had a degree of orthomorphism [readability in terms of its good shape] part way between a modern Admiralty chart and a metro or underground schematic. It, together with Homer's writings, indicates the safest, most efficient way to cross the Atlantic to take advantage of prevailing winds and currents . . .

Even if they could not determine longitude, Lusby has illustrated that after an initial exploratory mission the Minoans would have been able to find their way to the sources of copper in Lake Superior and then navigate their way home by using latitude only.

||||||||||||||||||||

There is another reason why I feel that the Minoans were slightly more comfortable in calculating latitude than longitude and it has to do with a recent re-interpretation of the evidence on a tiny golden coin. After the collapse of Minoan power, the Phoenicians inherited the remains of their Mediterranean trading empire. There are many indications that the Phoenicians travelled to Iberia, Britain, Ireland, India, Africa and possibly even America. Did they inherit Minoan maps, I wondered, and if so could these maps be found? I searched for a long time with no success, then via our website a friend referred me to the work of Mark A. McMenamin, Professor of Geology at Mount Holyoke College, Massachusetts. A palaeontologist, geologist and celebrated fossil hunter, he is nevertheless a much published authority on the Phoenicians, their language, coins and maps – a very rare combination indeed.

Professor McMenamin has studied a number of coins minted in Carthage, the Phoenician western capital, between 350 and 320 BC. The provenance and authenticity of these coins has not been challenged. Of relevance to this story is a particular golden coin, on which a horse stands proudly on top of a number of symbols.

Scholars originally surmised that these symbols were letters in Phoenician script, a theory that was discounted in the 1960s. Following 3D imaging analysis of the coin, McMenamin has interpreted the design as a representation of the Mediterranean, surrounded by the land masses of Europe and Africa with, at the upper left, the British Isles. If he is right the Professor has shed a radical new light on the 'discovery' of the New World.

To the left of the Mediterranean, under the horse's left rear hoof, is what he believes is a depiction of the Americas. So McMenamin postulates that the Phoenicians reached America – which I am quite sure they did.

Latitudes on the McMenamin 'Phoenician' map are pretty good. The longitude of the Atlantic and of America, by contrast, is drastically foreshortened. This would be accounted for by the navigator determining longitude by dead reckoning – he would not have appreciated how far west he had travelled with the help of the current. The map would thus not show the true width of the Atlantic. Where I respectfully differ from Professor McMenamin is that I believe that the initial provenance of the map on the coin is Minoan.

My reasoning is prompted first of all by the locations displayed on the map. It details all of the places which Minoan fleets visited, including the British Isles, the Baltic and the Indian Ocean. In short, Professor McMenamin's map coin shows the Minoan trading empire. More importantly, there are some particularly Minoan aspects to the map – for instance the importance (from their size) of the representations of Crete and Cyprus. Most crucially, the Mississippi, which the Minoans followed to reach Lake Superior, appears on the coin's representation of America.

It seems to me that the initial information to compile this map came from Minoan sources: perhaps other maps which have since been lost.

In the light of all this new evidence, I think the Minoans may have had the capacity to use dead reckoning to draw up simple world maps. I believe an original Minoan map will one day be found and authenticated. It will show all of the places the Minoans visited – from the relatively straightforward seas of the Mediterranean, Crete,

Cyprus, the Middle East and Iberia, to the quite simply audacious: Ireland, Britain and the Baltic. Not to mention North America, Africa and India – destinations that took extraordinary levels of bravery and daring to reach.

My belief is based upon the very exact geophysical locations of the observatories that, in my opinion, the Minoans either built or adapted. These sites in Kerala (South India); Malta; Stonehenge; northwest France; Ireland; and the Orkneys – and on the Elbe and the banks of Lake Superior – span nearly half the world in longitude from South India (77 degrees east) to the Great Lakes (89 degrees west) – a total of 166 degrees. Moreover, the latitudes allow for the cross-checking of results – Babylon (32 degrees north); Malta (35 degrees north); Brest, northwest France and Lake Superior (both 48 degrees north). To have done all of this required planning; a sense of overview. In other words, it required maps.

||||||||||||||||||||||

By comparing lunar eclipses on the same day (achieved by counting sunrises) the Minoans could trace the moon's passage across the sky and its position relative to a fixed star. This would help them create ephemeris tables (records of the co-ordinates of celestial bodies at specific times) of the moon for Kerala, Babylon, Malta, Stonehenge, probably northwest France and, as I was soon to find out, Lake Superior. Having an observatory in America would make a lot of sense, because it is such a long way west of our Bronze Age meridian of zero degrees longitude – the magical datum line I believe the brilliant Minoan navigators set at Stonehenge. They did this so they could cross-check and refine results and extrapolate them to make ever more accurate ephemeris tables, in the same way that dec- lination tables could be made for each day, to enable latitude to be determined by using the sun. In short, they could make world star maps for the northern hemisphere from India to Lake Superior.

I asked the former Royal Navy Admiral Sir John Forster 'Sandy' Woodward for his view on my theory. Could the ancients really have achieved all of this? His thoughts were:

The whole business of going trans-ocean would have been very rough – enough to get you there but not all that much more. In fact, rather like my cross-channel voyages in a 21ft sail boat – I didn't bother much with accurate navigation. I headed well to one side of my destination [the uptide/upwind side] and turned downtide/downwind when I reached the coast until I reached the place I'd intended. OK, so a compass, the tide tables, etc., made my DR [dead reckoning] pretty good, but as I was keen to show with 'pool navigation', approximate navigation is usually entirely adequate. [Pool of Error navigation is discussed on my website.]

In a ship, you can never forget about obstacles: reefs and rocks, even icebergs. Even the ultra-modern *Titanic* was lost to an iceberg, four days into the ship's maiden voyage. The cruel seas took 1,517 lives.

This would have been a dangerous business, especially if the Minoans met with conditions of low or zero visibility, heavy rainfall or snow; mists and fogs. Yet copper and tin were the most valuable substances in the known world. Wouldn't the Minoans have risked life and limb to find them?

33

A METALLURGICAL MYSTERY

Over the past seven years, since we set up our website, we've had hundreds of emails from North American readers of *1421* and *1434*. They all tell a tale of a mysterious conundrum. To this day the perplexing story of America's missing copper is taught in American and Canadian schools. The story began with Professor Roy Drier, who in the early part of the 20th century was Professor of Metallurgy at the Michigan College of Mining and Technology. The mystery itself, however, dated from the Bronze Age.

In the 2nd millennium BC millions of pounds of copper were mined out of mineral-rich Lake Superior, in North America. Yet where are the Bronze Age artefacts to show for it? While Bronze Age relics do exist, there is a significant mismatch between the number of finds and the evidence left by the miners. The copper, and the bronze it helped create, appears to have vanished into thin air. Could the Chinese explorers I wrote about in *1421* have taken the copper ore back home to China, my correspondents asked?

I didn't have the answers, but I knew enough to start digging. I was also reminded by Dr Gunnar Thompson (see chapter 31) that some of the copper artefacts that are found in American burial mounds show evidence of foreign influence.

Immense wealth – in the form of gold, silver and amethysts – lies just beneath the surface of the vast body of water that is Lake Superior. What's more, Lake Superior's mines were the richest source of copper on earth. Over a billion years ago, copper crystallised in the lava bed that lay deep under the waters of the largest and most northerly of the Great Lakes of America. Glacial action

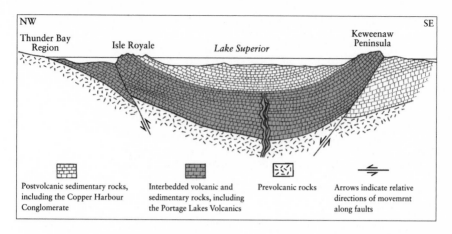

NW

SE

Thunder Bay Region   Isle Royale   Lake Superior   Keweenaw Peninsula

Postvolcanic sedimentary rocks, including the Copper Harbour Conglomerate

Interbedded volcanic and sedimentary rocks, including the Portage Lakes Volcanics

Prevolcanic rocks

Arrows indicate relative directions of movemrnt along faults

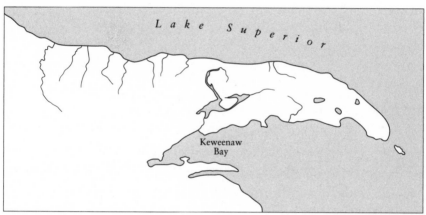

Lake Superior

Keweenaw Bay

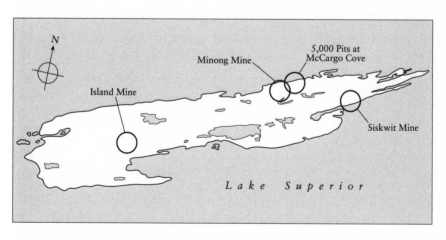

N

Minong Mine

5,000 Pits at McCargo Cove

Island Mine

Siskwit Mine

Lake Superior

273

exposed some of those mineral riches, in some cases leaving vast rocks of 'float copper' in river beds, on lake shores and glinting out of rocks. On Isle Royale, a particularly copper-rich island in the northwest of Lake Superior, and on the Keweenaw Peninsula, a further site on the lake's southern shore, Drier found over 5,000 mines. Those workings were not from our time: they dated from between 3000 and 1200 BC...

Drier wondered if the missing copper had been made into goods that were no longer in North America. In other words, he wondered if the resource had actually been exploited by outsiders and not by the native peoples of America. In the 1920s, so soon after the loss of the *Titanic*, this was an almost unthinkable proposition. Could seafarers from the Old World have achieved such a crossing? The mystery remained unsolved.

IIIIIIIIIIIIIIIIIII

The Menomonie Indians of north Wisconsin have a perplexing legend that speaks of an ancient network of mines. The stories describe the mines as being worked by 'light-skinned men', who were somehow able to identify the right place to dig by throwing magical stones on to the ground. When they struck, the instruments made the copper ores ring, like a bell. It's possible that the legend conflates the start of the mining process – finding the ores – with the end results of it; in other words, with creating a metal. A similar practice was used to find tin in Europe during the Bronze Age.

S.A. Barnett, the first archaeologist to study Aztalan, an archaeological site near the Menomonie Indians' native lands, believed that the miners of ancient times originated from Europe. His conclusion was largely based on the type of tools that had been found there, tools which he said were not used by the local people. But where was the workforce and where were the villages, the rubbish tips, the burials? The answer: nowhere. Only their tools remained.

The two world experts on this mystery, Octave Du Temple and Professor Drier, staged a number of expeditions to Isle Royale, hard on the Canadian border, in the 1950s. Their research gradually convinced them that an ancient civilisation far beyond North America had in fact mined and taken away the copper. At first they considered the Egypt of the pharaohs as a possibility, but they eliminated that line of research when they could find no evidence of Egyptian ships having reached North America.

They did, however, manage to fix dates for when the copper was mined. Charcoal found in the bottom of two pits on Isle Royale in Lake Superior confirmed that they had been worked from at least 2500 to 2000 BC.

A huge copper nugget found 5 metres (16.5 feet) down in the Minong mine on Isle Royale weighed 2,390 kilograms (5,270 pounds). There is an old black and white photograph of it in the Detroit Public Library. The copper had been hauled out of the ground on stretchers, or 'cribs'. In another pit, a crib of solid oak had somehow survived without rotting, because of the anaerobic conditions. It had been crafted out of the original oak tree over 3,000 years ago. Mining had certainly been going on here for many centuries before the birth of Christ – efficient mining, conducted on a huge scale.

The ancient miners built large bonfires on top of the copper-bearing veins. Once the rock was glowing with heat they split it by pouring cold lake water over it. Solid copper of extraordinary purity was taken from the veins and smashed with stone hammers. But there seemed to be no trace of these people. They left behind no carvings, writings or paintings. And they left no dead – or so it appeared. Their tools remained in the American mines, just as if the miners expected to return on the following day. They never did. They disappeared – into thin air, it seemed – around 1250 BC.

These ancient miners were clearly very highly skilled. Every major seam opened in the area had already been worked in pre-historic times. Carbon dating of the wood timbers left in the pits puts the first mines at 2450 BC, and the abrupt end of the mining at 1200 BC. The idea of the 'singing' device used by the miners to find the copper was strange, but there was no doubt: these people had been expert miners.

Even today, it is difficult to source original metal ore; in our own times, the price of copper is an economic bellwether, because it is used in so many things, from washing machines to houses. The estimates of the amount of copper ore excavated in the Lake Superior region thousands of years ago are staggering, although they vary hugely. One claim is that as much as 230 million kilograms (500 million pounds) of extremely pure copper was mined here. Another estimate is far more conservative, putting the total at 1.4 million kilograms (3 million pounds).

Yet whatever the precise figures, you would still expect to find large numbers of Bronze Age artefacts here in the New World, either as pure copper or as alloyed bronze. Not so. That copper – tons of it – has disappeared from the archaeological record. The estimate is that if all of the historic copper artefacts of the right date ever found in North America were added together, they would still account for less than 1 per cent of the copper mined from Lake Superior.

||||||||||||||||||||

We'd chosen the former French fur trading outpost of Thunder Bay as our base and had reserved seats on the port side of the connecting plane. On the beautiful, bright afternoon of our flight, we could see the island stretched out to the southwest, clear as crystal.

Although Isle Royale is technically in the United States, Thunder Bay is right on the border, on the western arm of Lake Superior. Isle Royale itself has been turned into a nature park and is now about as remote a wilderness as you can get.

We were up at dawn, eagerly reporting for duty outside a stout Edwardian classical revival building, the Brodie Resource Library, a minute before opening. Apparently, this part of Ontario holds the world record for wind chill factor: luckily, a helpful librarian let us in out of the cold.

Her name, as it turned out, was Wendy Woolsey and she gave us the quick five-minute tour, explaining that when the library opened in 1885 the Carnegie Foundation had wanted 'to promote a desire for good wholesome literature'. Along the way, the owners had made sure that readers could also get cleaned up: in the original building, the smoking and recreation room had contained a bath. It was good old, paternalistic stuff; during staff picnics, games were to be played and 'steps taken for the safety of women and children'. I noted that this effectively meant 'no liquor'.

The library staff, Wendy, Karen Craib and Michelle Paziuk, were extraordinarily helpful, and kindly agreed to trawl through the library archives looking for:

1. DNA reports on prehistoric miners' skeletons and the dating of these skeletons.
2. Prehistoric miners' skulls, especially those which had been carbon dated, with reports on their heritage – whether European or Native American.
3. Reports on the prehistoric mines, not least methods of mining and the tools left behind.
4. Rock art – particularly any pictures of prehistoric ships.
5. Descriptions of artefacts found by the first Europeans to reach Lake Superior – not least by the Jesuits.
6. Legends and folklore of the indigenous Native American peoples. In particular whether they had mined or assisted in mining in prehistoric times.
7. Chemical analyses of Lake Superior copper and in particular Isle Royale copper. Was it really 99 per cent pure? Were there any corroborative chemical analyses?

The five of us had an amazing day. Marcella and I discovered that even today Thunder Bay is a major port, with extensive grain storage facilities. Isle Royale is 15 miles (24 kilometres) from here. The huge poking finger of the Keweenaw Peninsula, another major site known for ancient mining, is a further 40 miles (64 kilometres) south of us, across the water on the Michigan side of the vast Lake Superior. An official 'water trail' has been mapped out for those who want to visit what was once called 'Copper Island', but it would take a full five to ten days to complete, depending on weather conditions.

Meanwhile, our three helpers were unstoppable. A list of the reports they helped us dig up is on the website.

While Karen dredged up reports on mining, Michelle tracked down skeletal remains and local myths and Wendy looked up local art finds. She also guided us through textbooks and papers that contained chemical analyses of the local ores.

The mining methods Karen's reports described are identical to those used at the Great Orme in north Wales. The Bronze Age metal hunters smashed out the ore with copper axes, adzes and awls which were almost straight duplicates of those at the Great Orme, and they bashed the excavated ore with stone 'eggs', just as they had done in Wales. Perhaps this was the device behind the folklore, behind the Menomonie Indians' myth? The 'eggs' were once again identical in shape, size and weight to those in the Welsh mine.

Thirteen chemical analyses of Great Lakes copper since 1894 at: the Kearsarge and Tamarack mines; Isle Royale; the Phoenix Mine; the Quincy Mine (Keller); the Quincy Mine (Ledoux); the Atlantic Mine; the Osceola and Franklin Mines; Lake Superior (Carpenter 1914); Lake Superior (US Bureau of Standards 1925); Keweenaw (Phillips 1925); and Lake Superior (Voce 1948), show trace elements in the copper of 0.09 per cent or less. That is, the copper is 99 per cent or more pure. The ten copper ingots found in the Uluburun wreck (out of over 300) as analysed by Professor Hauptmann and colleagues can only be Lake Superior copper, because of this extraordinary purity.

Karen also found page after page of detailed descriptions of prehistoric pottery and tools from thousands of prehistoric mines. We looked at old mine workings; mauls and hammers; skeletons and skulls; copper spears and arrows; knives; chisels; punches; awls; needles; harpoons and fish hooks; necklaces; spatulas. There was even a 'Minoan-style' double-headed axe.

The excavation was in 1924 and was known at the time as the 'Milwaukee Expedition'. Over 10,000 Bronze Age artefacts taken from the mines are now in the Milwaukee Public Museum. Experts like Professor N.H. Winchell later argued that the indigenous peoples had no knowledge of refined metalworking. He also had intriguing evidence that the prehistoric miners had had a distinctive genetic characteristic: a remarkable flattening of the shinbone.[12] That took me aback a second – didn't the 'Amesbury Archer', or 'King of Stonehenge', and his 'son' both have a marked bone peculiarity?

Michelle had so far had no luck with ancient skeletons. She did, though, find us some highly interesting information on the changes in the water levels of the Great Lakes over the past five millennia. The fact that the water levels had changed radically since the Bronze Age would later become a crucial piece of evidence in my search for the truth.[13]

The first modern-day mining operations began near the Ontonagon River on the Keweenaw Peninsula. Wendy had dug out details of a strange, so-called 'perched' rock at Pequaming which is carved with a prehistoric Caucasoid human face. The rock is aligned with a dolmen that sits on top of Huron Mountain – and is in line with sunrise at the winter solstice. Could it have been used by the Minoans as a kind of signpost? Like a child in a sweet shop, I hardly knew which morsel of information to chew on first.

34

ADVENTURES BY WATER

The imposing scale of Lake Superior comes as a surprise, rather like the sight of the Grand Canyon for the first time. The colours of the shoreline are vivid and varied: red and grey granite, white quartzite, black basalt and golden beaches. The hills behind the shoreline soar to a height of 305 metres (1,000 feet). This is the southern end of the Canadian Shield, a 4-billion-year-old almost soil-free area of the earth's crust, with its edges clawed but barely diminished by the grumbling glaciers.

And to the west, nothing. No land, no ships, no sails, no people. Just water.

Isle Royale was turned into a national park in the 1950s and today it certainly qualifies as a wilderness – 22 miles (35 kilometres) from Grand Portage, Minnesota; 56 miles (90 kilometres) from Copper Harbor, Michigan; 73 miles (117 kilometres) from Houghton, where the boat service starts. Six and a half hours by boat and then a long hike from the camp site through wild country.

The largest island on the largest freshwater lake in the world, Isle Royale has no modern infrastructure to speak of – no bridges, causeways, or even roads. A deserted, marshy island in the middle of nowhere and populated, I'd been told, mainly by moose. Oh, and wolves.

This is the middle of mining country. You can see on maps that the lake's silting-up has changed the settlement's position relative to the water. On a satellite image I could just make out the outline of a likely area, huge and flat, through sun-dappled shadows in the trees.

Why did these miners leave their tools so suddenly, as though they were planning to come back again the next day? Could there have been an epidemic? Could the whole community have fallen sick and died? Or was it connected with matters at home – the Thera eruption? I suppose that is something we will never be able to work out for sure . . .

IIIIIIIIIIIIIIIIIII

Another day; another island. This time on Lake Michigan, the mid-section of the great butterfly shape drawn by the Great Lakes on the face of America. Beaver Island – 45 degrees 39 minutes north/ 85 degrees 33 minutes west – is also known as 'the Emerald Isle'. It's a fertile gem of a place, set among the best fishing fields of the lake. Fourteen miles (22 kilometres) long, it is a thirty minute ferry ride from the city of Charlevoix.

Tracking the Minoans across the lakes was one of the most thrilling moments of my discovery. A treasure trail of clues had already been leading me towards this island, whose geographical position is key to its importance. A large number of storage pits that still hold traces of corroded copper lead from the Keweenaw Peninsula to Beaver Island.

That was as nothing to the news I had read of a stone circle on the island. Apparently, the native Americans call it a 'sun circle'.

A native American elder had told Professor James Scherz that a mystical series of stone circles lay submerged in the northern reaches of Lake Michigan. The elder told Scherz that these stone structures were all linked by what he called 'Thunderbird lines'. They all led to a large stone circle on Beaver Island.

In the 1950s Scherz made a study of the ring, which is made up of 39 stones and is 121 metres (397 feet) in diameter. He concluded that it was built for astrological purposes.

As my time in Canada drew to an end, evidence of a Minoan presence here was pouring in. North of L'Anse, on the Pequaming Peninsula in Keweenaw Bay, we'd found raised stone cairns that were probably used as beacon markers to guide in the ships. Archaeologists have found the remains of prehistoric cemeteries, created for the mine workers of old, near Green Bay opposite Beaver Island. Sea shells from the Gulf of Mexico and the North Atlantic lie beside the fragments of bone – in some cases, alongside copper jewellery. These copper artefacts appear to be identical in style to those found in the Uluburun wreck.

Hundreds of large, cast copper axe heads had been found – a hoard, just like in Europe – by an archaeologist named Warren K. Moorehead. Near Copper Harbor, above what would have been a beach 3,000 years ago after the retreat of the last glacial ice age, a petroglyph (rock engraving) of an ancient sailing ship. The design, roughly drawn, shows a ship that looks just like the graceful Minoan ships of Thera.

Most excitingly, a second stone circle has now been found nearby, near Traverse City: the only problem being that the site is now under water. Mark Holley, Professor of Underwater Archaeology at Northwestern Michigan College, discovered the series of stones 12 metres (40 feet) below the surface of Lake Michigan.

By his account, amidst the watery gloom Holley thought he could see that one of the stones appeared to have been carved. The markings are about 1 metre high and 1.5 metres long (3 feet by 5 feet). They are worn and the pictures Holley brought back were inconclusive: people speculate that the stone carries a pictogram of a mastodon (elephant-like mammal). Yet mastodon, say archaeologists, were not common this far north, and by this time the huge, tusked mammals were dying out.

The Minoans, I was learning, left little to chance. I admired their strategic choice of location. Beaver Island is about halfway between the southern shore of the narrow part of Michigan's Upper Peninsula and the entrance to Grand Traverse Bay. In other words, it is easy to get to with ships under your command and it's in a great strategic position. It would also be easy to defend – a great site for a trading centre.

Here, at the heart of America, there appears to be a Minoan star observatory, a mini-Stonehenge. Far from leaving without trace, the ancient mariners had left behind something infinitely more precious, to me at least: a mass of copper tools and artefacts. I felt closer than ever to these fearless seafarers, the sailors who had tamed the Atlantic Ocean to become the men of Atlantis.

35

A HEAVY LOAD INDEED

There was, however, a problem to resolve. How would the Minoans have transported the copper? There are only two possible routes: the first is east, through the Great Lakes. You would have to work your way laboriously across Lake Erie and Lake Ontario to the St Lawrence and then head for Newfoundland. As a sailing challenge this is very, very hard to do. The advantage would be that you would be on the right latitude to cross the North Atlantic quickly, to reach bases in Britain. The problem is, it's highly dangerous.

No one is sure whether Giovanni Caboto (John Cabot), the first European since the Vikings to reach the eastern coast of Canada, landed at Newfoundland or Cape Breton Island in 1497. Aided by strong west winds and the Gulf Stream, he took only a triumphant fifteen days to get back. However, it almost goes without saying that the North Atlantic is never to be trusted. In 1498 Cabot tried to do it again, but he never made it. He and his five ships, commissioned by the ambitious English King Henry VII to find 'the Indes', were lost at sea.

The other, safer, route could be via the mighty Mississippi, the largest river system in North America. It is less than a mile from some parts of the southwestern shore of Lake Michigan to rivers that are part of the Mississippi watershed. There are also a number of places in Wisconsin, south Michigan and north Indiana where you could carry boats (in a process known as 'portage') from the Lake Michigan watershed to the Mississippi watershed. Looking at the map, I could see that such an approach would take the Minoans south via evocatively named places like Poverty Point and Cahokia.

Would journeying this far south have made sense?

Yet again, Wendy and her colleagues were being wonderful, supplying me with a lot of papers. It looked to me as if, due to geographic change, this route south had been navigable in the Bronze Age in a way it isn't today. I've made a note of which reference sources were the most useful in the search later in this chapter, but I am starting here with Professor James Scherz, the first man to examine Beaver Island's stone circle. He had also made a special study of ancient trade routes.[14]

Professor Scherz writes:

Immediately after the glaciers melted, water levels of the Great Lakes were much lower than today with the main outlet through North Bay [i.e. into the St Lawrence]. But as the land rose under the melted glacier, the river at North Bay also rose. So did lake levels behind it, until the waters of Lakes Huron, Michigan and Superior combined into a giant body of water called 'Lake Nipissing' ... but the water could continue rising only so long, and finally a southern outlet opened into the Illinois River over the present Chicago Ship Canal [at the southern end of Lake Michigan. 'CSC'].

Another book showed me pictographs of very old sailing ships – beside Lake Superior. Of course. When Lake Superior and Lake Michigan were one, there was no water level difference at Sault Sainte Marie. So there I had my answer: the Minoans would have been able to ship their precious copper out of America by sailing the grandest river of them all. The Mississippi.

I was still unsure. How on earth, I kept thinking, did they get upriver against the current? And then I realised. I'd got it wrong: the bones were all there, but I'd assembled the skeleton wrongly. Of course, they didn't; they waited for a favourable wind to be able to sail against the current. Large vessels from the Mississippi could then have sailed directly into Lake Nipissing, and then on to Keweenaw and Isle Royale. And coming back, they would have used the current – the current on the Mighty Miss can be very strong in high water periods and the speeds range from 1 to almost

6 knots. If they caught the tide right, travel was dead easy.

If copper was floated south, then there should be evidence of that downriver: evidence of smelting going on, for instance. Finding that would be my next task. It would have been relatively easy to float the copper south on rafts – as indeed logs, cattle and corn were transported south in the early days of European settlement.

The Mississippi would have connected the Great Lakes with the Gulf of Mexico and thence the Atlantic. The river is a series of wide streams – for the most part a steady, fairly even water flow. Sailing north against this flow is tedious, but it is possible in high summer, when the prevailing wind is from the southwest. Mark Twain describes sailing on the Mississippi against the current and some Spanish explorers also used the river to travel north, as did the French Jesuits. The Minoans also had experience of sailing upstream on the Nile – a river which is just as long as the Mississippi.

A bit of further reading and I had calculated that it would take around eight weeks to get from the Gulf of Mexico to the Great Lakes, sailing and rowing against the current. Floating south, on the other hand, was simple, using the current as the method of propulsion.

HOW WAS IT DONE?

The mining areas are shown on the map 'The Great Lakes' map at the front of this book. Various mines are marked with crosses – please also refer to the diagram showing the Geology of Michigan copper in Chapter 33, and James Scherz's 'Ancient Trade Routes' in America's Copper Country' (*Ancient American*, issue 35). At the time those mines were operated (2400–1200 BC) the ice sheets had retreated, to leave Lake Superior ice-free in summer. Mining would take place during the summer months and the ore was carried by ship or raft across the Great Lakes, which were then distinct, but connected, sheets of water. Come winter it would only be possible to move the copper across the lakes by sledge or raft. Mining would not have been possible, given the severe cold.

TRANSPORTATION – LAKE AND RIVER SYSTEMS

The most northerly river, the Ontonagon, flows north through the copper mines into Lake Superior, while the others – the Wisconsin and Rock Rivers – flow south into the Mississippi. As a consequence copper from all of the principal mines could be shipped downstream to Lake Superior using the Ontonagon River or the rivers on Isle Royale which connect the mining area to Lake Superior. This explains the loading harbours found on the northern part of the Keweenaw Peninsula (Pequaming, Anse and Baraga) and around Lake Superior (Otterhead harbour and on Isle Royale). So using natural resources Minoan miners could collect copper and then ship it to Lake Superior without going against the current or negotiating rapids.

LIVING QUARTERS

Come autumn, Lake Superior, the surrounding rivers and the land would have frozen over. In summer, miners would need living quarters within reach of the mines. The map of the Great Lakes shows the areas in which towns fortified sites have been found.

THE VOYAGE SOUTH

I drew my finger southwards down the map, trying to take in the names of the many places at which the Minoans may have rested. It is a long way to what is now Louisiana: they must have stopped to trade what they could for food, and they may well have stopped to process the copper.

'Are there any major prehistoric sites you know of that could have acted as a trading centre?' I asked Wendy, who was fast becoming a near-expert on the Bronze Age.

She knew of several, but the most interesting was a major early native Indian settlement at the mouth of the Mississippi. It was a name I had noticed before on the map: Poverty Point. There had even been claims that in times past this very site was Atlantis: might that be a folk memory of the people who had once come here to trade?

Whether or not it had a link to the real Atlantis, this intriguing ancient site had definitely been a major trading point. It had been built over a long period, between 1650 and 700 BC – eight centuries after the building of the Great Pyramid. The dates worked. So did the fact that huge firing ovens have been found near there – as had copper from the Great Lakes.

FOOD

One would have imagined maize would not have grown this far north in 2000 BC, but surprisingly the crop has been found in human graves at Baraga on Lake Superior. It may have been brought up the Mississippi – as I will describe later. Fish and game would have been abundant. Lack of vitamins would have otherwise been a severe problem in winter as would have been lack of potable water; perhaps deep wells were in use and these were not frozen.

A CAMP TO TRADE NEAR POVERTY POINT

On the Pearl River mouth of the Mississippi, near where it debouches into the Gulf of Mexico, we find one of the most revealing bits of evidence. The Claiborne and Cedarland Rings, very near to the trading centre of Poverty Point – and contemporaneous with it – sat on high ground above the marsh created as the Mississippi debouches into the Gulf of Mexico. Excavated by James Bruseth of Louisiana State University in the 1970s, the sites had unfortunately been quite extensively damaged by relic-seekers.

Bruseth was called in just before the bulldozers arrived: a new harbour facility was being built, which was how the site had been uncovered. Along with large concentrations of charcoal, the archaeologist found 'an enormous hearth, as long as a football field: 6 foot [1.8 metres] deep and 300 feet [91 metres] long'. The huge 91-

metre hearth was on a level with a number of smaller hearths, between 50 and 60 centimetres (20 and 26 inches) wide. He put a radiocarbon date of 1425–1400 BC on his finds. He also found large numbers of clay moulds: presumably the remains of the moulds used to cast the copper into ingots.

In the middens, or rubbish pits, Bruseth found hundreds of bits of broken clay which he thought must have been thrown away because they had been broken during the firing process. The fact that they were broken tells us quite the opposite. These were clay moulds, broken when the Minoans hammered them apart to get at the precious cast metal inside. Ingots of pure copper.

How did the Minoans get to Poverty Point?

36

INTO THE DEEP UNKNOWN

This is what I think happened. Early Minoan explorers crossed the Atlantic with the Equatorial Current, the 'free ride' that would be relatively easy in summer before the hurricane season. They visited Meso-America; the proof that they traded cotton, fruit and vegetables is discussed by Professor Sorenson, who has tracked these exchanges.[15]

After visiting Yucatán, the current would have carried them north to the Mississippi Delta, where they discovered that today's 'land of opportunity' was even then a land of unimaginable riches. They saw float copper (copper lying loose in the soil) being used by the people of Poverty Point, who were international traders. At Poverty Point they were told of the original source of the copper in the Great Lakes: particularly at Lake Superior. On reaching the Great Lakes they saw huge nuggets of pure copper simply lying on the ground. They built a stone observatory on Beaver Island to fix the latitude and longitude of this incredible treasure and drew up a map. Then they set up an entire system to bring the riches back to Europe – creating protected townships for the miners in summer at Lac Vieux Desert, and winter quarters at Aztalan and perhaps Rock Lake.

Lac Vieux Desert is part of the drainage shore of the Mississippi today. Sometime between 1300 and 1200 BC the Aztalan settlement was abandoned, for reasons that remain unknown to this day. During the Bronze Age both the Lac Vieux Desert site and Aztalan were fortified, as well as being protected by water – in the middle of a lake (Vieux Desert) or by the Rock River (Aztalan). Recently, walls and substantial pyramid structures have been found sub-merged in Rock Lake. Both sites are protected by double embank-ments. The miners clearly feared attacks – by humans or bears.

I assume Lac Vieux Desert was the summer camp, from which mines could be reached downstream via the Ontonagon River. Aztalan, further south and therefore warmer, was the winter quar-ters, reached via the Rock River from Lake Michigan.

They built shipyards on Isle Royale, at Otterhead Island and at the northern end of the Keweenaw Peninsula at Pequaming, Baraga and L'Anse. Later they had the sad duty of constructing burial mounds for those workers who died at Aztalan, Rock Lake and Green Bay.

The copper was loaded on to pallets, which were extended to form rafts. These rafts were floated down the Mississippi via Lake Michigan to Poverty Point in the south. Here the vast kilns awaited them. The rafts were broken up to be used for charcoal: the copper was turned into ingots before they'd even left, and stored to await collection by Minoan ships which crossed the Atlantic.

On the return leg, the ancient traders pitched up just north of the Mississippi Delta – and waited for Nature's largesse. Who knows? Perhaps it was Nature herself who taught them how to get home. We know that 19th-century whalers understood about the currents, because we have records of them watching the great humpback whales hitch a free ride north, and then following them. Our intrepid Minoans may well have followed the loggerhead turtles, fabulous creatures born under a million stars on the beaches in Crete, as they began their epic journey home to breed.

On the return journey they again found a free ride on the Gulf Stream. Warmed by the sun in the shallow waters of the Gulf of Mexico, water rushes east through its only point of escape, the narrow Straits of Florida. Thirty million cubic metres (39 million cubic yards) per second of water drain out of the Florida Straits, pumping our Minoan ships north.

It's difficult fully to grasp the immensity of this great ocean river. To carry just the sea salt that flies through the Straits every hour would take more ships than exist in the entire world today. A billion cubic feet of water streams past Miami every second of the day.

The push through the straits, only 50 miles (80 kilometres) wide and 2,500 feet (760 metres) deep in places, increases the speed and force of the Florida current. Deflected to the north by the Bahamas, the Florida Current then joins with the Antilles Current. The Gulf Stream System, as it has just become, then triples in volume and pushes north.

The Minoans may have whiled the hours away watching the whales, as the huge mammals fed off the plankton-rich waters along the current's edge. Or they could have seen the blue sharks that use the current to reach their pupping grounds off southern Ireland, west Wales, Spain and Portugal. The total round trip made by these amazing predators is 9,500 miles (15,300 kilometres).

The current flows north along the southeastern United States, slowing as it runs toward Cape Hatteras, and then turns towards the east. Off the Grand Banks of Newfoundland, the Gulf Stream and the Labrador currents collide, creating fogs and storms famous for their treacherous nature. The ocean water temperature changes are often dramatic: as much as 20 degrees of change as you cross from one current to the other.

The huge wheel of the North Atlantic continues to turn. As they pass Nova Scotia the water flow increases to 150 million cubic metres (195 million cubic yards) per second, pushing them on to the Outer Hebrides (Lewis) and the Orkney Islands. Here at their bases their compatriots have erected 'observatories' – stone circles as at Stonehenge – where they can compare and update their star charts. They then follow what would later become the English, French and Spanish coasts as they set out for home.

Now the pole star is sinking lower in the heavens. When it reaches the latitude of the Pillars of Hercules, they know they must head due east – and prepare for a joyous return to Crete.

I could rest now, happy that I had solved my conundrum. The solution to the mystery of the missing copper, I now realised, had been staring me in the face at the beautiful, autumnal Lake Superior: in the form of Mother Nature.

Nature provided everything – from the copper sitting right on the surface, ready to be collected, to the wood for the rafts and the water to carry them. Mother Nature was behind the significant change in water levels over three millennia. And in riding the Gulf Stream back to Britain, the Minoans used what came perfectly naturally to them – the power of the seas.

37

SO: THE PROOF

I decided to look in detail at the prehistoric mining tools and copper implements found on the Keweenaw Peninsula and on Isle Royale, starting with the stone hammers.

The fact that Great Orme and Isle Royale copper miners used identical mining tools could, of course, be a coincidence. It could simply be put down to the designer's favourite adage: form follows function.

What about the other copper implements left by the miners for other purposes? To ascertain the full extent of the similarities we built up a file of prehistoric Lake Superior copper implements.

We also took photographs at a number of American museums and obtained others from an excellent website called *The Great Lakes Copper Culture*. Thousands of different copper and bronze tools have been found around Lake Superior: projectiles with flat, conical and rat-tail points and square and ornate sockets; harpoons with flat and round tangs; knives of every shape and function – with crescent-shaped, straight and curved blades; all manner of fish hooks; scrapers and spatulas; axes, adzes, palstaves; punching instruments in the form of awls, needles, gorges, mandrils, finger drills; fasteners in the shape of staples, clasps and rivets; personal adornments – bracelets, rings, beads, gorgets, tinkling earrings. There were literally hundreds of varieties of copper and bronze implements, running into thousands of individual items.

Then it struck me. If, as I had been contending, it was the Minoans who had mined the copper of the Great Lakes, then we should compare artefacts like for like with artefacts found in Crete, on Thera and in the Uluburun wreck. So I re-examined the museum catalogues. These, time and again, showed photographs of bronze utensils identical to their Lake Superior counterparts of the same age. You can view a selection of compansons in the colour plate section and on our website.

IIIIIIIIIIIIIIIIIIII

DETERMINING THE SOURCE OF COPPER ARTEFACTS THROUGH TRACE ELEMENT PATTERNS AND X-RAY SPECTROMETRY

A number of American geologists agree that Great Lakes copper was of extraordinary purity. Professor James B. Griffin of the University of Michigan puts the total trace elements in the material at 0.1 per cent or less: i.e. the copper is of 99 per cent purity or more.

It should be possible to use non-invasive X-ray spectrometry to measure the trace elements in copper artefacts from Lake Superior and the Uluburun wreck and then use the results to discover if the Great Lakes were the original source of the copper.

As far as I know, two attempts to use this method have been made, in studies thirty years apart. The first one was made by Edward J. Olsen[16] and the second by Georg (Rip) Rapp, James Allert, Vanda Vitali, Zhichun Jing and Eiler Henrickson.[17]

I found both easy to read. Essentially they both reached the same conclusions. There are so many variables that any conclusion reached has to be on a 'balance of possibility' basis (my wording) and treated with great caution. It may one day be possible to be certain, but much more work is required. George Rapp and colleagues state that their purpose was to (1) indicate the extent to which sources of natural copper can be chemically distinguished; (2) present a methodology for trace-element sourcing; (3) publish a small database; (4) provide another means by which archaeologists can approach complexities in trade and exchange networks. There is a final caution concerning finding a 'reasonable' geographic source area. (As soon as the word 'reasonable' appears, it seems to me that the possible area of the source becomes too wide to draw definite conclusions.)

Edward J. Olsen lays out the problems one after another. He explains lucidly how the method works. Each trace element in the copper sample is energised by X-rays and then emits a signal. The intensity of the signal denotes the strength of the trace element. The method can be used for trace element 22 (titanium) upwards. However the accuracy depends on:

1. Size of sample and whether it is an aggregation or one single mass.
2. The particular trace element and its relationship with copper: i.e. zinc's trace element 'peak' is so close to copper's 'peak' that it can be masked; another example, silicon, increases the apparent intensity of aluminium. Then the efficiency of the X-ray tube depends on the trace element being sought and how this differs from trace element to trace element.

He summarises:

Thus, in order to determine the chemical characteristics of copper [artefacts] from a given locality one must be aware that trace elements are exceedingly unreliable.

Bearing these two reports in mind, I wanted to double-check my contention that Bronze Age copper of 99 per cent purity was only found in the mines of Lake Superior. Unfortunately, the distinguished Professor Hauptmann believes it is impossible to be this precise. As he wrote to me: 'you cannot prove [10 Uluburun ingots] came from Lake Superior'.

Despite this I maintain my assertion – firstly because the reports assembled by Professor Griffin are not dealing with artefacts but with mined copper. Secondly, there is an almost complete lack of trace elements in the reports Griffin has assembled – which record the analysis of Lake Superior copper. We are not considering minute differences between one trace element and another but the virtual absence of all of the trace elements. The argument is about statistics, factorial probability, and not about the accuracy of the X-ray method deployed.

As to the actual artefacts themselves, I undertook an extensive study of my own. My first job was to separate out the different implements on the Uluburun wreck, then combine those objects with those found on the Seytan Deresi (16th century BC) and Cape Gelidonya (late 13th century BC) wrecks; the aim being to show exactly what was in use in the Mediterranean between 1600 and 1200 BC. All three wrecks were discovered off the coast of Turkey.

Then I decided to combine the implements from these shipwrecks with similar Bronze Age implements, tools and artefacts found at Minoan sites of the same age. These sites were in Crete, Santorini (Thera), some of the Cyclades islands and Mycenae. Marcella and I had already visited the principal museums in those places – the Heraklion and Knossos museums in Crete, the Archaeological Museum of ancient Thera; the National Archaeological Museum in Athens and museums in Mycenae and Tiryns.

The purpose of this exercise was to see whether any items found in the wrecks were different from those that could have been found in Minoan bases near and around Crete. As far as we could see there were none. The style of tools, weapons and implements found in the Uluburun and Cape Gelidonya wrecks were also found in Crete, Thera and Mycenae. In other words this was a check and balance to confirm that it was Minoan bronze artefacts, as opposed to various imports, that filled the wrecks. Images of the artefacts – separated out into those found in the two wrecks and four museums – are shown on our website.

COMPARISON BETWEEN MINOAN BRONZE AGE ARTE-FACTS FOUND IN CRETE, THERA AND THE BRONZE AGE WRECKS WITH THOSE FOUND IN LAKE SUPERIOR

We trawled through the Bronze Age bronze, copper and tin artefacts left by the miners who mined the copper from Lake Superior, in particular those from the Keweenaw Peninsula and Isle Royale. We separated these Bronze Age artefacts into the same categories as those from the Old World, as described in this chapter. As far as I can see, every item found in Lake Superior has its near-counterpart in Minoan artefacts of the time. Lake Superior's ancient miners had the same array of implements, weapons, tools and domestic equipment as the people who lived on Bronze Age Crete, Thera and Mycenae.

This could, of course, be simply the phenomenon known as parallel development. The Isle Royale miners needed bronze weapons to defend themselves, to hunt and to eat the food they caught. So they designed implements whose form would 'follow function', as a modern-day designer would put it. The logical result of that process, of refining a design until it reaches an optimum point, could end in tools that are highly similar to their Old World, Minoan counterparts. This argument will have its supporters, naturally. Yet how can it be coincidence that the measuring weights used so many thousands of years ago take the form of animals? (See second colour plate section.)

It is also intriguing that even local Native American myth appears to support the idea that it was outsiders who were mining the area's islands and peninsulas. This has a bearing on the other potent argument for the involvement of outsiders in Lake Superior's mining heritage: namely, the industrial scale on which the copper was mined and processed. This area was so rich in minerals that local people didn't need to mine: they could use the abundant float copper found on the surface. Those who mined the copper were clearly not indigenous Americans. I suggest the best explanation is that the miners were the same people as those who so efficiently controlled the bronze trade in the Old World.

The stone astronomical structures were so like those in Britain. Grave barrows, astronomical alignments, avenues and *cursi*: the elements were the same. One could just about argue that these basic structures, particularly those based on fundamental geometrical elements like the circle, could have been arrived at by different cultures acting independently of each other. Yet combine that with the discovery that the tools are of the same design and I felt that the evidence for a common culture was building. I knew that it could be a long haul to prove my case definitively, but it was getting there . . .

It does not of course follow that after a certain period in prehistory the Minoans used Lake Superior copper exclusively. But it is an inescapable fact that certain ingots in the Uluburun wreck do not match any known European source.

If I am right, we have another fixed date to work with: the Minoans were sailing to Lake Superior before the Uluburun was wrecked in 1310 BC.

Notes to Book V

1. J. Walter Graham, 'The Minoan Unit of Length and Minoan Palace Planning', *American Journal of Archaeology*, 64 (1960)
2. Cyrus Gordon, *Forgotten Scripts*, Basic Books, 1982
3. Anthony Johnson, *Solving Stonehenge: The New Key to an Ancient Enigma*, Thames & Hudson, 2008
4. Dava Sobel, *Longitude*, Fourth Estate, 1998
5. J. Fermor and J. M. Steele, 'The Design of Babylonian Waterclocks: Astronomical and Experimental Analysis', *Centaurus*, 42, 2000 pp. 210–222
6. J. M. Steele and F. R. Stephenson, 'Lunar eclipse times predicted by the Babylonians', *Journal for the History of Astronomy*, 28 (1997)
7. J. M. Steele, 'The Accuracy of Eclipse Times Measured by the Babylonians', *Journal for the History of Astronomy*, 28 (1997)
8. N. M. Swerdlow, *The Babylonian Theory of the Planets*, Princeton University Press, 1998

9. J. M. Steele, *Journal of the American Oriental Society* 119 (1991), p. 696

10. *Nature*, vol. 468, (2010) pp. 496–498

11. Homer, *Odyssey*, trans. Samuel Butler, 1998

12. N. H. Winchell, 'Ancient Copper Mines of Isle Royale', *Popular Science Monthly*, vol. 19, 1881

13. James B. Griffin (ed.), *Lake Superior Copper and the Indians – Miscellaneous Studies of Great Lakes Prehistory*, University of Michigan, 1961

14. Based on a number of research papers published in *The Ancient American – Archaeology of the Americas Before Columbus*, and by researchers from the Ancient Artefact Preservation Society, notably an article by Emeritus Professor James Scherz titled 'Ancient Trade Routes in America's Copper Country' (*Ancient American*, issue 35)

15. Professor Sorenson (Refer to select bibliography in this book)

16. Edward J. Olsen, 'Copper Artefact Analysis with the X-ray Spectrometer, *American Antiquity*, vol. 28, no. 2 (October, 1962)

17. George Rapp et al., 'Determining Geological Sources of Artefact Copper: Source Characterisation using Trace Element Patterns', *American Antiquity*, vol. 68, no. 2 (April 2003)

BOOK VI

||||||||||||||||||||||||||||||||||||

THE LEGACY

38

THE SPOTS MARKED 'X'

By now, I was running to catch up. I'd got solid proof that the Minoans had travelled throughout most of Europe and that they had explored a large part of North America. Now I needed to find out if there were any studies that showed a common thread. Tracking down DNA evidence would be my next line of enquiry.

'X' usually marks the spot in any treasure hunt. By pure coincidence, the same was true of my own hunt for gold – in the form of information. There is a widespread, but much contested theory that the first Amerindians originally came from East Asia. However, an intriguing DNA haplogroup has recently been found in several Native American populations – haplogroup X. The X group is the only Amerindian haplogroup that does not show a strong connection to East Asia.

What is interesting is that although haplogroup 'X is itself rare, it has a perplexingly wide geographic range: despite its relative scarcity the group is found throughout Europe and the Middle East.

In North America this haplogroup is found particularly among Native Americans, especially tribes living in and around the Great Lakes, while in Scandinavia, for instance, it is found in only 0.9 per cent of the population.

Haplogroup X2 is a rare subgroup of X that appears to have expanded quite widely in the Mediterranean and the Caucasus around 21,000 years ago. It's now more concentrated in the Mediterranean, particularly in Greece, along with Georgia, the Orkney Islands and the Druze community in Israel. I looked for a summary. Here is the best I found, written by Jeff Lindsay. As he points out, some geneticists believe that 'Lineage X' suggests a 'definite' – if ancient – link between Eurasians and Native Americans.

ENTER HAPLOGROUP X

The team, led by Emory Researchers Michael Brown and Douglas Wallace, were searching for the source population of a puzzling marker known as X. This marker is found at low frequencies throughout modern Native Americans and has also turned up in the remains of ancient Americans. Identified as a unique suite of genetic variations, X is found on the DNA in the cellular organelle called the mitochondrion, which is inherited only from the mother . . .

. . . Haplogroup X was different. It was spotted in a small number of European populations. So the Emory group set out to explore the marker's source. They analysed blood samples from Native American, European and Asian populations and reviewed published studies.

'We fully expected to find it in Asia, like the other four Native American markers [A, B, C and D],' says Brown.

To our surprise, haplogroup X was only confirmed in a smattering of living people in Europe and Asia Minor including Indians, Finns and certain Israelis. The team's review of published MtDNA sequences suggested it may also be in Turks, Bulgarians and Spaniards. But Brown's search has yet to find haplogroup X in any Asian population.

'It's not in Tibet, Mongolia, South East Asia or Northeast Asia,' [Theodore] Schurr told the meeting. 'The only time you pick it up is when you move west into Eurasia.'

Haplogroup X is found in several places outside of Asia, including among the Finns for example (Finnila et al. 2001), who are often thought to be an earlier group in Europe in the light of Y chromosome studies but nevertheless appear to share many MtDNA lineages with other Europeans. Detailed information about the mutations separating the X haplogroup from the Cambridge Reference and other European haplogroups are provided by Finnila et al. (2001): especially see their Figure 2.[1]

In chapter 7, I'd discovered that the Minoans most likely arrived in Crete from central and eastern Anatolia. One would therefore expect to find a high incidence of X2 from that area. In fact, the statistics live up to expectations. The figures in brackets are percentages.

Turks (4.4); Iranians (3.0); Nogays (4.2); Adygeis (2.5); Abazins (6.3); Kumyks (3.6); South Caucasians (4.3); Georgians (7.6); Armenians (2.6); Azeris (4.2).

Here is where X2 has been found in significant amounts:

Belgium (FTDNA 3)
Crete (Reidla et al)
Egypt (Kujanova)
Finland (Mishmar and Moilanen)
France (FTDNA 4)
Israel/Lebanon (Shlush)
Morocco (Maca Meyer)
Navajo (Mishmar)
Ojibwa/Chippewa – Great Lakes (Fagundes, Achilli, Pirego)
Orkneys (Hartmann)
Portugal (Pereira)
Sardinia (Fraumene)
Tunisia (Costa)

The geographical spread of X2 was most helpfully summarised by Finnila and colleagues.[2] It is replicated on our website.

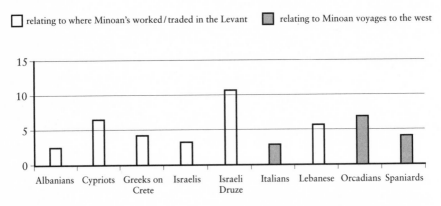

☐ relating to where Minoan's worked / traded in the Levant   ▨ relating to Minoan voyages to the west

The American inheritors of X2 (Ojibwa/Chippewa, around the Great Lakes) have their own sub-divisions of X2: X2a1; X2a2; X2aib; X2a1a; X2g.

From the higher percentages involved, it seems that X2 originated in the Near East, and particularly eastern Anatolia, just as the Minoans had done. There is much controversy over the dating of the DNA, i.e. when the mutations which resulted in the sub-haplogroups occurred. It would be difficult to say with certainty when the European carriers of X2 reached America. At present there are two possible dates for the arrival of X2: 9,200–9,400 years ago or 2,300–3,800 years ago. I would summarise the research in this way:

1. The implications are that the X2 populations in North America could have been caused by Y-DNA mutations from European X2 ancestors (c.1800 BC) or from a much earlier migration by Europeans to North America 9,420 years ago.

2. The people with one of the highest incidences of X2, the Ojibwa, live mainly around Lake Superior in Michigan, Wisconsin, Minnesota, North Dakota and Ontario (Canada) – i.e. near the copper mines described in chapters 33–37.

3. No fewer than sixteen European and Mediterranean countries in which significant levels of X2 DNA have been found were visited by the Minoans, as described in chapters 1–15.

4. DNA X2 has thus been found on both sides of the Atlantic and throughout the Minoan trading empire. Taking all of the evidence in the round, it appears that the Minoans could well have been the carriers of X2: it has been found where the Minoans originated, where they settled and where they traded. Other unknown European travellers appear to have reached America 6,000 years before them in substantial numbers.

5. X2 is not found outside the areas I have already identified as probable trading outposts of the Minoan trading empire.

Bearing in mind that X2 is also found where Minoan Linear A script is found, it seems to me that we can learn lessons from the DNA reports, not least:

(i) The Minoans originated from northeast rather than central Anatolia – because of the high incidence of X2 in that area and in neighbouring Georgia, Armenia, Iran and Azerbaijan (Finnila).

(ii) The mutation from X2 to X2a in the Ojibwa/Chippewa probably took place between 1800 and 1200 BC.

(iii) There seems to have been little sexual contact between the Minoans and the English (0.9) or the French (0.8) compared to that which took place between the Minoans and the Orcadians (7.2) or the Spanish (4.2).

(iv) The Minoans did not get to central, southeast or East Asia nor to West or southern Africa, but did apparently reach the Gulf States (Oman 1.3), Saudi Arabia (1.5) and Kuwait (2.0).

(v) The Minoan presence in what are now Israel, Syria, Lebanon and Jordan was even more pronounced, as evidenced by X2 in populations of Israeli Druze (11.1), Israelis (3.4) and Lebanese (5.8).

39

A NEW BEGINNING

In some ways this book was ending as it had begun. I had been working flat out, starting at 05:00 a.m. and working through for twelve hours each day. My right wrist was locked with arthritis and once again I was dog-tired. Still, my team and friends, especially Cedric Bell and Ian Hudson, had also been working hard, collating a stream of evidence from the well-wishers contributing to our website. I owed it to them to finish the manuscript and get the book published.

Nonetheless, the prospect of describing the end of the Minoan civilisation filled me with foreboding. For the Minoans, the end of empire was a slow and painful death, not a sudden execution. I could hardly bear to think of it. All that brilliance, all that invention – and all their cultural and technological expertise lost.

Thera's volcano had given several weeks' notice of its anger, so it must have seemed as if disaster had been averted. The island was thoroughly evacuated: the absence of carbonated corpses, as found at Pompeii, tells us that. The real tragedy was yet to come.

For safety's sake, everybody on Thera most likely sailed to the mother island, Crete. How could they know what would happen next? For days there would have been grumblings, maybe even the flash of an odd explosion, coming from the direction of Thera. Then it happened for real. An almighty thunder clap seemed to split the horizon. Next came a sinister rush of sound, as ash and gases blasted up from the volcano's heart.

After the staggering eruption, the heavy sky must have felt leaden and threatening. A spreading stain of purple on the northern horizon heralded the sulphurous clouds to come. I imagine a terrible moment of calm; a charged silence as, inexplicably, the sea reared back from the shore, leaving a mass of gasping and struggling sea creatures in its wake. Then came a low and distant roar. A gathering wall of water raced across the sea towards the mother island, its sheer force pulverizing everything in its path.

A hundred-foot wall of water hit Mochlos, the harbour town on eastern Crete. The tsunami probably destroyed the entire Minoan fleet, as it lay tied up in the various ports. If the first surge didn't pull them apart, the next one would have seen to that. The evacuees recently arrived from Thera were all sheltering in the coastal towns, so the disaster may have taken many lives. A superheated pyroclastic surge of gas, light ash and pumice may have followed. At Mochlos, it probably killed everyone. Plato's words came back to me:

> In a single day and night of misfortune, all your warlike men in a body sank into the earth, and the island of Atlantis in like manner disappeared – into the depths of the sea.

In Crete the human casualties, the destruction of houses and temples and, above all, the total annihilation of the shipping on which the island's prosperity relied must have been crippling. After the volcano came days, possibly weeks of toxic gas and lethal ash falling thick upon the ground, suffocating people and plants and poisoning the water supplies. The catastrophe appeared to be coming from the very skies the Minoans had studied for so long. They must have thought the gods had abandoned them. There may still have been plenty of people left alive. But the towering wall of water – and the smothering ash – had destroyed everything of the Cretan harvest's famous bounty.

Then came famine.

IIIIIIIIIIIIIIIIIII

A telephone call. Would I be interviewed by Greek national TV about the new book I was thinking of doing? It was a great excuse to get away. Marcella and I decided, as we had done on a previous Christmas, to revive morale with a special trip.

We drove to Venice and from there we boarded a magnificent car ferry, which at noon sails splendidly past St Mark's Square and into the Adriatic, *en route* to Greece. Forty-three years ago we sailed this same route, sleeping on the upper deck. Now, in old age, we can afford a cabin. As night falls we dine on rabbit cooked in white wine, gazing at the gathering darkness over a calm sea.

Six hundred years ago it was Venice that controlled the seaways of the Mediterranean. Her fleets were based at the (Croatian) island of Hvar in summer and further south at Corfu in winter. Pirates were ruthlessly crushed, free trade made safe. Three thousand years earlier, the Minoan fleets had performed the same role.

Marcella and I were sailing in the wake of history.

Once Minoan control of the seas collapsed, the pirates that King Minos had suppressed would have had free rein. The Adriatic is peppered with hundreds of small islands, perfect for camouflage and shelter. There are many more such in the Aegean.

The Minoans' extraordinary trade network did not immediately waste into nothing. For two hundred years or so, from around 1400 to 1200 BC, it seems that Mycenae took over the controlling role of the seas. The great city-state also assumed power over Crete. We don't know for certain whether this was the action of an ally helpfully stepping in to fill a power vacuum, or whether this was a hostile takeover by an aggressive force. When Mycenae's influence eventually waned, the era of safe seas ended – and so did the glories of an age.

Quite suddenly, between 1225 and 1175 BC, the Bronze Age ended in the eastern Mediterranean. So did the voyages to northern Europe and the Americas. The Keweenaw and Isle Royale mines on and around Lake Superior stopped production at least as early as 1200 BC. English tin mining stopped at the same time, as did work at the Great Orme copper mine. Bronze Age settlements in La Mancha, southeast Spain, were all abandoned too.

Many reasons have been advanced for the sudden collapse of the

Bronze Age – a crisis in civilisation that some have put down to comets, climate change, earthquakes, sunspots, or plagues. A great disaster had befallen the sophisticated world of the eastern Mediterranean. Some scholars call it simply 'the Catastrophe'.

Most scholars now agree that the Catastrophe was caused by a mysterious new military force. They were a people who left few traces behind them, save fifty years'-worth of widespread destruction. Their ferocious armies fought with new-fangled weapons such as javelins and wore defensive armour such as greaves and corselets. Their small, round shields suggest that they may have evolved radical new battle techniques, using infantry soldiers to great effect. The Egyptians called them 'the Sea Peoples': to this day no one knows who they were, or where they came from.

Between 1225 and 1175 BC the raiders overran the great civilisations of Crete, Mycenae, Anatolia and Upper Mesopotamia. Many of the cities and fortified palaces along the eastern coast of the Mediterranean, from proud Mycenae through Troy to the ancient Anatolian kingdoms of Kode and Hatti, were sacked. The Sea Peoples swept through the Amorite city-state of Emar and the city of Ugarit – both were in what is today known as Syria – and may even have reached as far inland as Hazor, north of the Sea of Galilee. As Robert Drews writes:

> The catastrophe peaked in the 1180s BC and ended about 1179 BC, during the reign of Ramesses III in Egypt, virtually the last of the great Pharaohs. The regimes in the region had been stable, palace-centred, wealthy and relatively peaceful. What followed, at least in Greece, was a Dark Age.[3]

Two hundred and fifty years after the Minoan civilisation on Crete had collapsed, the Hittite and Mycenaean empires fell. All in all, forty-four cities were lost to the Sea Peoples – until finally Ramesses III managed to defeat the marauders. Lower Mesopotamia and Egypt escaped the wholesale destruction. Yet both were fatally weakened, and a two-decade famine in Egypt all but destroyed that remarkable empire.

The explosion on Thera was the first mortal wound to the Bronze Age cultures of the eastern Mediterranean. Then the Sea Peoples came along and delivered the *coup de grâce*. The very thought depressed me. What would have happened to the world had its first great sea power, the ancient Mediterranean's brightest jewel, not been destroyed?

40

A RETURN TO CRETE

I had to keep my appointment with Greek national television. The interview itself passed off without incident. Following me in the studio discussion, although I didn't know it, was a charming and unassuming man from Crete. I have relied on his inspired research more than once: his name cropped up when discussing Minoan mathematics (chapter 32). Dr Minas Tsikritsis was on the programme to talk about his ground-breaking research into Linear A, one of the world's greatest linguistic conundrums.

The next day, Marcella checked our emails. 'You've got a new message. From a Dr Minas Tsikritsis.' Marcella and I were still treating our stay in Athens as a bit of a vacation, but holidays were soon to be off the agenda. I looked at the message, first curiously, then in shock.

Dr Tsikritsis had been studying my old friend, the Phaestos Disc. Not only that, but he felt that we would have a lot to talk about in terms of Minoan astronomy. Apologising for his use of English, the professor said he wanted to talk to me about his specially written computer programme. It had helped him to decipher some key aspects of Linear A – would I like to know more?

Dr Tsikritsis was excited. But not as excited as me. I read on. His new translation of Linear A, he said, allowed him to identify Minoan text and inscriptions wherever they could be found. He had identified inscriptions in Linear A written on ancient stone markers found by explorers in hundreds of different locations, which proved that the Minoans had visited India, the Baltic, northern Europe and Greenland.

Many of the ancient Minoan writings he'd decrypted tallied with the archaeologists' current view, that the entire Aegean had been under Minoan sway. The empire was, in effect, a confederacy of twenty-two cities. But, he said, that was not all: the Linear A evidence had convinced him that the Minoans had founded further colonies, in (here I copy his email direct):

1. Sicily
2. Syria, Palestine
3. The Bosphorus
4. Bavaria and Baltic Sea. [They travel for obtaining amber.]
5. Greenland for obtaining pewter [tin and lead]
6. Nordisland
7. India (a colony called 'Asteroysia')
8. Arabian Gulf. [There are findings of writing in caves in a Minoan colony in Paghaia Island.]

'I have evidence (photos, bibliographic references in Greek literature) for the above findings.'

His research matched my own findings – despite being arrived at through an entirely different process. Giddily, I did a quick calculation. With nine locations involved, the chance of this being a coincidence was factorial 9,360,000 to one.

⁞⁞⁞⁞⁞⁞⁞⁞⁞⁞⁞⁞⁞⁞⁞⁞⁞

Lovely though Athens was, this lead was more important. Marcella and I hurried to get back to Crete. This time, we were headed for the very centre of the island, atop a precipitous hill at Skalani, outside the lovely restored village of Archanes.

Dr Tsikritsis, his wife Chryssoula and his son Dimitris greeted us with the kind, warm hospitality that Cretans are famed for. He is a thoughtful man with strong features and even stronger opinions.

The family took us into the garden and showed us their organic olive trees, their wood oven and the tomatoes growing underneath the vines; an organic way of keeping the pests away. Tsikritsis gave

315

me a bottle of his lemon-infused home-pressed olive oil, made to a traditional recipe that his father, also an ancient history enthusiast, had rediscovered and handed down to him.

'You are the only one to count the oars,' he said to me, rather unexpectedly, referring to the Thera frescoes. 'This is twenty-eight oars each side, a big, big ship . . .'.

Then we entered a light, airy drawing room, built on two levels. In one corner was Dr Tsikritsis' study, a small, book-lined cubby hole with a telescope and a traditional Cretan wooden trestle stool. We hunkered down for a long talk.

As he began to explain his methods – with Chryssoula and Dimitris occasionally translating – it became clear to me that this was Dr Tsikritsis' life's work. He initially learned ancient Greek from his father, an expert in ancient scripts. Later, he developed his own specialist skills with degrees in mathematics and physics. He also has a master's degree in the methodology of religion and a doctorate in content analysis from the Aristotle University of Thessaloniki.

Few linguistic experts have a similarly broad skill base. For twenty years Dr Tsikritsis had devoted every moment of his spare time to understanding Minoan culture and deciphering Linear A.

Tsikritsis explained that to help him break the hieroglyphic code behind Linear A, he'd used the technique of consonantal comparison. It was crucial that he was already familiar with ancient Greek, Cypro-Minoan and the Mycenaean script Linear B, as well as with Cretan hieroglyphs. However, it was his more modern skills – in computing and mathematics – that had made the crucial difference. The breakthroughs started to happen when he tried statistical cryptographic techniques. He used many different Minoan texts as the basis for the work. Unfortunately for scholars, to translate a language with absolute statistical certainty you need at least fifty-six symbols. The Phaestos Disc has only forty-eight. He has been busily finding other tablets and artefacts to help him go further with the translation and it's this breadth of knowledge that seems to have been key. He told me what had inspired him.

'I think my first inspiration was when looking at a spiral design on a ring. The Minoans used spiral designs all the time, just as on the Phaestos Disc. This ring, I suddenly realised, could be read backwards, as well as forwards.

'... Another breakthrough was realising that there were fifteen symbols that were identical to Linear B.

'... Then I realised that a symbol's meaning could be changed by the word you put it next to.'

This was where the contextual analysis came in.

This is all complex stuff, but as Dr Tsikritsis showed me chart after chart of comparisons I saw it all unfold as a rational system: a beautiful, flowing language. It was simply astounding, the amount of evidence: again and again, on tablets and discs several millennia old, Dr Tsikritsis' systematic solution to the ancient mystery of Linear A seemed rational, consistent and clear.

His translations, time and again, involve that most magical substance: bronze. The Minoans accorded the alloy a special significance; almost a reverence. The translations show the Minoan society's huge prosperity and document the vast amounts of grain, pottery, olive oil and other goods that they shipped across the world. Dr Tsikritsis has even unearthed documents that show that this extraordinary society distributed food and goods to each according to their need. He also showed me the photographs of the ancient inscriptions he had mentioned in his email, which he would identify as being written in Minoan Linear A. They are found from Norway to the Arabian Gulf.

One of his most astounding discoveries is about the Minoans' in-depth understanding of mathematics, which helped them develop their knowledge of the stars.

'In mathematics it's always thought that the Babylonians and the Egyptians were far more advanced than the Minoans,' he said. 'But in 1965, Mario Pope found something unique – a proper fraction. It was written on a wall in Hagia Triada.'

Hagia Triada, just 2.5 miles (4 kilometres) west of Phaestos, was a town with a royal villa at its centre. The inscription, which reads $1{:}1\frac{1}{2}{:}\ 2\frac{1}{4}{:}\ 3\frac{3}{8}$, shows each number progressing by one and a half times the previous one. The calculation may have been drawn casually upon the wall for working out interest payments. What's striking, said Dr Tsikritsis, is that we know that the Egyptians also studied mathematical progression. They, on the other hand, only used integers. This formula was mathematically far more sophisticated.

Once he first began to suspect that the Minoans' grasp of mathematics was as inventive as that of the Babylonians, Dr Tsikritsis made it his business to discover other formulae that the Minoans had left behind. He discovered that they could count; in tens of thousands if they needed to. They could add, divide and subtract. What was interesting was their need for large numbers: the large amounts of goods and grain they were trading demanded that they developed this skill.

Showing me many diagrams, he convinced me that the Minoan use of geometry was unrivalled. For instance, to construct their signature spiral designs, they had to understand the use of tangents and cosines. It is Archimedes who is famed for defining the spiral in *On Spirals*; yet Archimedes' account didn't emerge until around 225 BC.

One of Dr Tsikritsis' more electrifying ideas could be really highly controversial. He is convinced that almost every single Minoan ceremonial object or building conforms to the 'golden mean' of φ, or 'Phi'. One of the most talked about, disputed and revered aspects of practical mathematics used in the arts and architecture, Phi is also known as 'The Golden Section,' or 'The Golden Ratio'.

This is a most significant claim. You can express Phi, the golden ratio of divine proportion, through this equation:

$$\varphi = \frac{1+\sqrt{5}}{2} \approx 1.6180339887\dots$$

Dr Tsikritsis has found this specific proportion of 1.61 in literally hundreds of Minoan objects and buildings he has measured and he does not think this can possibly be a coincidence. Today the Greeks usually attribute the discovery of Phi to Pythagoras, who lived c. 570–495 BC.

One ancient Minoan object in particular illustrates his point about Phi and the art of proportion, on two levels. It is an exquisite stone vase, found at the lesser-known palace of Zakros, that has what appear to be mysterious fire scorch marks on it. The Heraklion Museum dates it to 1500–1450 BC, although Professor Tsikritsis believes it may be older.

It is a libation vase, also known as a *rhyton*, which was used for ceremonial drinking. On it, you can see a design – a shrine or a temple, shown in a mountain landscape. Now brown with age and smoke, this precious thing was originally much prized. It is decorated with gold leaf. Both the proportions of the vase itself and the design inscribed on it conform to 'the golden ratio'. Draw a rectangle over the vase as a whole or the image of the shrine and then measure it: you will find that the proportions conform to Phi, says Dr Tsikritsis. In other words, both the form of the vase itself and the imagery upon it are composed using a mathematical ratio that we have always attributed to the ancient Greeks – and not to the Minoans. These are the very proportions that Iktinos and Kallikrates later used to build that greatest of world temples, the Parthenon, creating an unrivalled sense of harmony and serenity. 'This,' Dr Tsikritsis said, warming to his theme, 'is all about beauty.'

All of his studies show that the Minoans believed that Phi's beauty and harmony was sanctified. A vase constructed according to the Golden Mean would, they believed, be holy. Its very perfection would purify the water: a bit like the idea of Feng Shui today.

But it was when it came to describing the Minoans' love of the stars that these new revelations really came alive for me. Here I felt I knew these fascinating people. Dr Tsikritsis' theory is that for the Minoans the constellations were not merely stars. They felt they were gods, who lived and moved in the sky. Not only that, but, much like the Chinese believed until very recently, the ancient Minoans were convinced that their ancestors had joined those gods in the skies. They had become celestial bodies.

One of the absolute masterpieces of Minoan gold work, the Isopata ring, illustrates Dr Tsikritsis' ideas perfectly. The entire ring is only 2 centimetres (three-quarters of an inch) across and the skill it must have taken to cut it is extraordinary. It was found in a tomb at Isopata, near Knossos. (See second colour plate section.)

Four women seem to be enjoying an ecstatic ritual dance; their heads, though, are not human. They have the nodding heads of wheat or corn. In the background you can see the symbols of an eye and a snake. A smaller figure drifts downwards into the picture, as if she is from far away: she may be a goddess descending from the heavens. Dr Tsikritsis' theory is that the snake, which pops up time and again from object to object, is a Minoan symbol for the constellation of the Corona Borealis. In English, we call it the Blaze Star, but it is also known as the 'Northern Crown'. The crown, he believes, was given to King Minos' daughter Ariadne at her wedding ... the Corona *is* Ariadne, forever guiding her people out of the maze.

This ring, Dr Tsikritsis believes, is a type of sacred calendar. It shows a way of counting down towards the rainy season, using the position of the stars ... and the symbolic know-how of Ariadne. The stars didn't just guide the Minoans and inform them about the changing seasons – the Minoans thought that their ancestors were now their guiding stars: celestial bodies.

||||||||||||||||||||||

Dr Tsikritsis and I agreed that this complex and detailed subject was definitely meat and drink for another book. In the meantime, we all needed lunch. We'd walked through the village, reached the stone-built restaurant and ordered our food, but still we couldn't stop swapping notes. There was so much to talk about.

'What about the Phaestos Disc?' I asked.

'I haven't translated it all,' Dr Tsikritsis was keen to stress. 'I think that at least one side is a ... Τραγούδι.' I must have looked blank. He turned to Chryssoula. They had a quick discussion about the translation.

'A *tragoudi*,' said Chryssoula, leaving me none the wiser.

'... A sailor's song,' said Dimitris.

'... A sea-shanty,' I murmured.

I could scarcely believe it. Sitting in this warm and cloistered room, with the age-old aroma of slow-roasting lamb drifting into our nostrils, all of us seemed suddenly closer to the ancient world than to the modern one. I remembered the happy scenes of the frescoes, those sailors arriving back in Thera, the people thronging to the shore to greet them. The Phaestos Disc, a fire-hardened clay roundel found in the charred remains of a ruined palace, was what had started me out on my quest. Why had the mysterious object struck me quite so strongly? It seemed so appropriate, somehow. The disc that had fascinated me so much records a departing sailor's sea-shanty.

41

THE LEGACY OF HOPE

After this adventure Marcella and I returned to Athens with a lot more energy. Before meeting Dr Minas Tsikritsis, I had pretty much decided that most of the Minoans' fantastic cultural legacy had been entirely lost. The idea had filled me with gloom. Now I could see that was simply not true. The Minoans may have suffered terribly, but their work, their invention and even their sense of fairness had lived on.

In the fading sun we wandered at the foot of the Acropolis, admiring the play of light and shade along the Parthenon's soaring columns. The temple, designed to meet the ultimate standard of perfection, the Golden Section, points proudly towards the Bay of Salamis. Here Themistocles' fleet destroyed the invading Persians' mighty navy. The Parthenon is still to me the most beautiful building in the world: now I understand why.

> In this island there existed a confederation of kings of great and marvellous power, who held sway over the island, and over many other islands also.[4]

Plato's words echoed in my mind. The Minoans had achieved exactly this. Like ancient Crete with its mysterious Linear A, Plato's Atlantis was a literate state; we know this because he describes the god Poseidon giving out rules, which the first prince 'inscribed on a pillar of orichalcum'. What was 'orichalcum'? It was a copper alloy. Plato describes such a substance lining the walls that entered the city of Atlantis, which gradually became more and more lavish as you approached the main temple.

322

... And they covered with brass, as though with a plaster, all the circumference of the wall which surrounded the outermost circle; and all that of the inner one they covered with tin; and that which encompassing the acropolis itself with orichalcum, which sparkled like fire ... [5]

and a little later:

... All the exterior of the temple they covered with silver, save only the pinnacles and these they covered with gold. As to the exterior, they made the roof all of ivory, variegated with gold and silver and orichalcum.[6]

The Minoans were master smiths in metal and, of course, they were fabulously wealthy: it was certainly conceivable that they could line their city walls with decorative panels of bronze, copper, silver and gold. I realised now that everywhere that I had been, Plato's majestic concept of Atlantis had dogged my every footstep.

As I'd realised in chapter 3, Plato's text suggests that the ancient metropolis and the Royal City were separate entities. This bears a strong resemblance to the relationship between Crete and Thera. The main city, he said, lay on a circular island about 12 miles (19 kilometres) wide. The Royal City, meanwhile, lay on a rectangular-shaped island. So Plato's Atlantis was certainly two islands and possibly more. There are plenty of islands to choose from within the *Pax Minoica*. Plato claimed that the kings of his fabled Atlantis had 1,200 ships; as I've explored in chapters 6 and 19, Crete had certainly had ships – in their many hundreds. Plato believed that Atlantis had suffered an environmental crisis and that the soil had become depleted. This is exactly what I'd found out had applied to the whole region of the eastern Mediterranean. Not only that, but he'd said that the people of Atlantis had been brave enough to breach the Pillars of Hercules. All of these things the Minoans had done, and more.

As Plato says, in those far-off days the ocean was navigable; since there was in front of the strait which I've heard you say your countrymen call 'the pillars of Heracles'. This island was bigger than both Libya and Asia combined; and travellers in those days used it to cross to the other islands, from where they had access to the whole mainland on the other side which surrounds that genuine sea.[7]

The sea that we have here, lying within the mouth just mentioned, is evidently a basin with a narrow entrance; what lies beyond is a real ocean, and the land surrounding it may rightly be called, in the fullest and truest sense, a continent.[8]

This, with thanks to a new translation by Rodney Castleden, can only be America, across the Atlantic. Plato writes about Poseidon:

Poseidon ... named all his sons. To the eldest, the king, he gave the name from which the names of the whole island and of the ocean are derived – that is, the ocean was called the Atlantic, because the name of the first king was Atlas.[9]

Most crucially, Poseidon and Cleito were the parents of five sets of twin brothers and they divided the island equally between them. Those brothers and their descendants had founded not an island, but an entire empire. I could see it now. What I was confronting was not a 'lost island', but the Lost Empire of Atlantis.

Clearly the great classical tradition hadn't all begun here, in ancient Athens. Iktinos and Kallikrates' great temple was built on a great legacy. The golden era of classical Greece had evolved from a heroic tradition of enterprise and adventure. It was the fortunate heir to a much, much older civilisation. With new eyes I could now trace a clear evolution from the purity and grace I'd seen in Minoan architecture – a delicacy of form that came about due to the Minoans' love of spiritual and mathematical perfection – to the eloquent classical ideal of the Golden Number. From there on, the genius of Greek architecture had flourished. It went on through revival after revival, affecting everything from Renaissance Rome to 18th-century Washington. The Minoans' remarkable influence did not simply stop dead in its tracks.

There were also the 'inventions' that we have always attributed to the glory that was classical Greece: coinage; a system of standard weights; music, architecture and art; theatre; even the very idea of spectacle. As Arthur Evans had pointed out, the long-robed Cretan priests of Ayia Triadá were playing the seven-stringed lyre a full ten centuries before it was supposed to have been invented on the island of Lesbos.

Perhaps the greatest legacy is the idea of art for art's sake – and the pursuit of knowledge for its own sake. The Minoans gave the world exquisite painting, ceramics and jewellery and an appreciation of the finer things in life. Their building technology was superb, their ideals were glittering. They did have rulers, but they believed in share and share alike: out of this generous impulse, revolutionary ideals such as democracy would eventually take shape.

This legacy also applied to war. When the Greeks won the crucial battle of Salamis against the invading Persians they had behind them the inspiration given them by a long-standing tradition of shipbuilding technology: it was a baton that had been handed them by the Minoans.

ATHENIAN TRIREMES

The Athenian ships deployed at Salamis were called triremes. They were on average about 5.5 metres wide and 39.5 metres long (18 feet by 130 feet), approximately the same size as the Minoan ships before 1450 BC. Like the Minoan ships they were dual-purpose vessels, with a large square sail on a horizontal yard when in trading configuration and with the mast stepped and powered by oars when in military mode.

The Athenians had modified the Minoan ships in such a way that two banks of oarsmen, one on top of the other, were used when fighting. These newer ships carried 150 men rather than the Minoan ships' 120, so they could be rowed faster than the Minoan ships, to reach a speed of 10 knots. Their principal weapon was a ram in the bows, to pierce the enemy's hull. However, they were inferior to the Minoan ships when they were underway, with a higher centre of gravity that made them much more likely to capsize in a gale in the open sea.

Just as the Athenian ships were developments of the Minoan ones, so was Athenian weaponry and armour. The Athenians' helmets and shields were made of bronze, as was their body armour – one can see the similarities in the bronze Minoan armour that has been found in Crete itself and in Mycenaean graves.

At first sight the idea of militarism fits badly with the Minoans' reputation as a carefree society which did not need the protection of soldiers and fleets – after all, Crete's palaces were not guarded. Yet recent research by Stuart Manning has shown that although Crete itself was not protected from invasion, having no defensive walls to speak of, the wider empire was. The further away from Crete, the greater the level of protection. At the height of Minoan power and influence in the Aegean, there were no fortifications on the 'Minoanised' islands closest to Crete – that is Kythera, Thera and Rhodes. However the more distant sites, such as Ayia Irini on Keos and Kolonna on Aegina, were fortified. Mycenae's great defensive walls are explained by the fact that it was on the mainland, and less easily defended.

The Minoan lead in art, science and astronomy was inherited by a Greek civilisation that went on to produce ageless works of art and literature; the Greeks invented theatre, devised calendars, became skilled engineers. Cherished ideals such as citizenship and democracy and disciplines such as philosophy and science followed.

Most astonishing of all is that the Minoans achieved all of this 2,000 years before the birth of Christ, 1,500 years before Buddha or Confucius and 2,500 years before Muhammad. Beauty had been this people's watchword. They had shown the world that a peaceful existence was a profitable one. They had rid the oceans of pirates and then with luck, daring and great maritime skill they had voyaged on adventures beyond imagination. The Minoans hadn't just been the bringers of bronze: they breathed the Bronze Age into life.

To me, the story of the Minoans – and as I now realised, that of the people of Atlantis – is truly one of wonder. But it is not one of fantasy. This was no underwater daydream. Yes, this was a society lost to history. But it was not a lost race of miraculous beings with fantastical powers. It was a real place of real achievement, where lived a people whose brilliance and resourcefulness resounds through the centuries. Atlantis was not one place, but an empire of many places – an empire that reached out across the world, bringing a magical new technology with it.

This is a tale that tells us one thing: that the history of this world is far more fascinating, complex and indeed more beautiful than we could ever imagine. Most important of all – what do you think?

Gavin Menzies

London

St Swithun's Day 2010

Notes to Book VI

1. *Science*, 1998, vol. 280, p. 520
2. Pubmed Central Table 1, *American Journal of Human Genetics*, 2003; November 73(5) 1178–1190 as Table 1
3. Robert Drews, *The End of the Bronze Age*, Princeton University Press, 1995
4. Plato, *Timaeus*, 25a, trans. Robin Waterfield, Oxford World Classics, 2008
5. Plato, *Critias*, 116b, trans. Robin Waterfield, Oxford World Classics, 2008
6. Plato, *Critias*, 116d, op. cit.
7. Plato, *Timaeus*, 24e, op. cit.
8. Plato, *Timaeaus*, 25a, trans. Rodney Castleden in *Atlantis Destroyed*
9. Plato, *Critias*

TIMELINE

|||||||||||||||||||||||||||||

B.C. 300,000
◇ Indo Europeans settle in central Anatolia.

B.C. 100,000–B.C. 5000
◇ Emigraton of these people to Greece and Crete, via Rhodes.

B.C. 4500
◇ First copper smelted in Crete.
◇ Almendres stone megalith observatory started (Portugal).

3200
◇ Minoans from Crete arrive in South East (Mediterranean).
◇ Spain found mining colonies (Millaren culture).

3000
◇ Millaren culture peaks c.2600 B.C.
◇ Minoan seals show ships with masts and sails.

2900
◇ Minoan seals and Egyptian scarabs of that date found in Crete.
◇ Tinto copper mine starts production SW Spain (Ortiz).

2800
◇ Minoan trading contracts with Syrian – Palestine coasts, Byblos, Ugarit, Mari flourish.

2650
◇ Saqqua Pyramid completed.
◇ Mari producing exquisite jewellery.

2570
◇ Khufu Pyramid completed.
◇ Huge demand for copper and tin.

2500
◇ Great Orme Copper Mine (U.K.) starts production.
◇ Copper axes found in U.K.
◇ Stonehenge Phase II (Sarsens) started.
◇ Avebury man.
◇ Malta invaded by sea peoples.
◇ Megalith observatories started at Villa Nova de San Pedro.
◇ Fortress built in estuary of Tagus (Portugal).

2450
◇ Indian port of Lothal in operation – visited by Harappan Traders (Rao) and Minoans.
◇ Stele showing Sumerian soldiers with copper armour, copper helmets and copper weapons.

2340
◇ Sargon I Emperor of Akkad – huge demand for bronze.
◇ Minoans in Britain for tin (Waddell and Rawlinson) Sargon attacks Crete? (3 times – Sargon autobiography).

c.2280–1930
◇ Stonehenge phase III completed.

2200
◇ Bronze axes appear in Britain (Needham).

2100
◇ Egyptian scarabs common in Crete.

2000
◇ (1950) 'Palace' trade between Crete and western group of Islands – Thera/ Melos, Kea. Protopalatial Cretan pottery at Akrotiri (Thera); Phylakopi (Melos);

Ayiarini (Kea) Cretan palaces are prime movers in long distance trade.

◇ Large quantities of Minoan pottery at Fayun Egypt, Kamares pottery at Byblos, Ugarit, Beirut, Qatna and Hazor. Laurion silver exported via Crete.

2030

◇ Ferriby ship (planks) built in Britain.

2040–1640

◇ Red Sea Nile Canal built.

◇ Cretan textiles to Egypt in substantial quantities (Buck). Portraits of Cretans in Egypt bringing gifts. (Buck).

◇ No evidence of direct trade between Levant and Egypt nor Cyclades and Egypt/Levant – Trade in Cretan ships (Buck).

2000

◇ Bronze age hoards found in estuary of Rio Minho (Portugal).

◇ Flat copper axes in use in Ireland.

1900

◇ Tod treasure found below temple near Luxor, contains Mesopotamian materials and silver cups from Crete, labelled with Amenemhet I (1922–1878) cartouche.

◇ Goods from Crete appear in Mari. King of Mari sends Cretan gifts to Hammurabi of Babylon. 'After the foundation of the Palaces, Crete became an international player as never before . . . a major presence on eastern Mediterranean stage'. (Fitton).

◇ 'Protopalatial period represented first great flowering of Minoan culture' (Fitton).

2400–1800

◇ Yemeni bronze hoard.

1800

◇ Trade between Crete and Keos, Delos, Thera, Naxos, Aegina, Kythera, Paros and Amorgos (Buck/Scholes). Minoan settlement in Aegina (Buck). No evidence of direct trade between Middle Helladic Greece and Egypt (Buck) – Goods carried in Cretan ships.

1783

◇ Tell el Dab'a founded.

◇ Minoan decorated palace. Minoan ships use harbour.

1785

◇ Indian cotton appears in American middens (Sorenson).

1700

◇ In MM III (1700–1600) Minoans engage in large construction projects to repair earthquake damages.

◇ 'In the period of the second palaces [1700–1400] the island of Crete was home to a remarkable civilisation. Characterised by flourishing palaces, urbanisation on a scale not seen elsewhere . . . era widely seen as apogee of Minoan civilisation' (Fitton).

1600

◇ Volos ship.

◇ Dover Bronze Age Boat.

1500

◇ c.1492–1458 Hatshepsut expeditions to Punt (Somalia).

1450

◇ Thera explodes. End of Minoan Civilisation.

1400

◇ Rekmires Tomb (Thebes) shows American corn (Thompson).

◇ Indian temples show American plants (Gupta).

EPILOGUE
PLATO AND ATLANTIS, THE LOST PARADISE

Listen then, Socrates, to a tale which, though passing strange, is yet wholly true, as Solon ... once upon a time declared.

This is Critias, a lone voice introducing a tale about a lost paradise, a magical Garden of Eden which was struck down by the awesome force of nature. This beautiful island was the cradle of civilisation, but was destroyed by the gods because of the arrogance, the hubris, of its people.

PLATO'S TALE OF 'ATLANTIS'

Long ago there existed an island, populated by a noble and powerful race. This beautiful place was the domain of Poseidon, god of the sea, who had fallen in love with a mortal woman, Cleito. He created a magnificent palace for her in the centre of the island. The people of this land possessed great wealth thanks to the abundant natural resources of the island, which was also a centre for trade and commerce. The rulers held sway not just over their own people but over the Mediterranean, Europe and North Africa.

For generations the people of the island(s) led a noble and unselfish life. They prospered from their skill in using copper and precious metals. But slowly, corrupted by avarice and greed, they changed. They decided to use their powerful navy to invade Greece and Egypt. Zeus noticed their immorality. He sent a huge wave. This drowned Atlantis, which vanished forever in a sea of mud. Greece was saved.

This is a brief summary of the tale related by Plato around 360 BC, in his dialogues *Timaeus* and *Critias*. These two accounts are the only known descriptions of Atlantis and have promoted controversy and debate for over 2,000 years. Many believe the stories are morality tales and fables, works of Plato's wonderful imagination. Others think that Plato may have been describing a lost civilisation, one that really existed, that he called 'Atlantis'.

FINDING THE TRUTH

It is extraordinary to think it, but the story of the Minoans is so incredibly ancient that even the ancient Greeks had forgotten it. History got lost in the mists of time. The tale was finally retold by Plato, but it is only because of this one sole author, and two texts, one of which is unfinished, that we know anything about ancient 'Atlantis' at all. So why, then, would we think it could possibly be true?

Here I have to acknowledge my debt to A.G. Galanopoulos, whose book *Atlantis: the Truth Behind the Legend*, was written with Edward Bacon in 1969. Together they mounted the first serious challenge to the orthodox academic view of the time – that Atlantis was total invention. It was Galanopoulos who first told the world the truth about the sheer scale of the 'Theran event', as the vast volcanic eruption is known. He was also the first to speculate that the tsunami that then hit Crete would have been of huge, destructive force.

It was also Galanopoulos who pointed out, quite rightly, the sheer number of times that Plato insisted that, although he was no historian, his account was based on truth. In Plato's two dialogues it is not just Critias who insists that the story is true: Socrates ends Critias' tale by saying:

> And the fact that it is not invented fable but a genuine history is all-important.

Plato makes the point that this is not 'a story', but historical fact not once, but four times. As Galanopoulos points out, Plato isn't creating a fictional world, the details of which were in his control. He actually seems worried about the inconsistencies in his account. For instance, he questions whether or not a trench as deep as he states it is could even be built. If this was indeed fiction, then why would he worry?

It seems fitting that these major breakthroughs should have been made by the top seismologist of his day. The climax of the story of Atlantis is also the story of one of the biggest geophysical events the world has ever seen. And with irreproachable scientific logic, Galanopoulos also worked out the solution to another of the great mysteries behind the Atlantis 'myth'.

Plato's account throws a few rotten eggs our way. He says that the date Atlantis was eaten up and buried under the sea was 9,000 years before the information was passed on by an Egyptian priest. Plato also greatly exaggerated the size of Crete, doubling its actual size. He gives the dimensions of the plain of the Royal City as 3,000 by 2,000 stades. Both figures have confused the picture. Scholars have triumphantly held them up to demolish the arguments in favour of Minoan Crete being Atlantis. It was Galanopoulos who pointed out the obvious.

'The solution of this riddle,' he said, 'is as simple as the mistake which created it.'

It was simply an error in the maths. Plato's Atlantis (Crete) is given as 3,000 stades, twice the length of Crete. Either Plato, or more likely the Egyptian priests who passed this information on, simply mis-translated the numbers.

As Bacon and Professor Galanopoulos pointed out, the parallels between Crete, Santorini/Thera and Atlantis are unavoidable. Minoan Crete was densely populated, as was Plato's Atlantis. Atlantis was divided into settlements each with a separate leader, but all subject to the Royal City. On Minoan Crete the king appears to have been the overall leader, with (let's call them) nobles governing other centres across the island in the king's name. The bull is crucial to

Minoan life and art; in the *Critias* (119c-120d) we find that this is the case in Atlantis, too:

> In the sacred precincts of Poseidon there were bulls at large; and the ten princes being alone by themselves, after praying to the God that they might capture a victim well-pleasing unto him, hunted after the bulls with staves and nooses, but no weapons of iron.

Plato is known to have visited Crete in person. What is not certain is whether he himself made the connection between 'Atlantis' and Crete. Here I have put together a commentary on some of the things Plato says, and how it is possible to interpret them, once you know something of Crete, Santorini, and their eventful pasts.

PLATO'S DESCRIPTION OF MINOAN CIVILISATION

Plato writes that the civilisation of Atlantis employed highly organised methods of agriculture. To cite *Critias*:

> ... It produced and brought to perfection all those sweet-scented stuffs which the earth produces now, whether made of roots or herbs or trees or of liquid gums derived from flowers or fruits ...

GM: Here Plato is referring to Crete's perfume industry in the Bronze Age, based on olive oil with terebinth resin as a fixative and perfumes of fruit and flowers (see chapters 8 and 10). Plato continues his description:

> ... The cultivated fruit also [vines] and the dry [corn] which serves us for our meals – the various species of which are comprehended under the name of 'vegetables' – and all the produce of trees which contains liquid and solid food and unguents and the fruit of the orchard tree so hard to store, which is grown for the sake of amusement and pleasure, and all the after-dinner fruits which we serve up as welcome remedies for the sufferer from repletion – all these that hallowed island [of Atlantis] as it lay beneath the sun produced in marvellous beauty and endless abundance ... [1]

GM: The 'hallowed island' of Crete provides everything Plato describes. Moreover Plato writes that the island is rectangular and that the island has heavy rainfall in winter, both true of Crete. Plato's island has mountains and plains in the same position as Crete's.

Plato states in *Critias* that the civilisation of Atlantis was a place of conscious amenity, leisure and public service:

> ... The springs they made use of, one kind being of cold, another of warm water were of abundant volume, and each kind was wonderfully well adapted for use because of its natural taste and these they surrounded with buildings and with plantations of trees such as suited the waters; and moreover they set reservoirs round about some under the open sky; and others under cover to supply hot baths in the winter; they put separate baths for the king, and for the private citizens, besides other women ...

GM: Phaestos and other Cretan palaces had all of these amenities (as described in chapter 1). By contrast, other great civilisations of the time in Egypt, the Levant and Mesopotamia had the amenities but were not islands.

Plato describes Atlantis as a literate state.

> ... The relations between her ten kings were governed by the precepts of Poseidon as handed down to them by the law and the records *inscribed* [my italics] by the first prince on a pillar of orichalcum which was placed within the temple of Poseidon in the centre of the island ...

GM: The Minoans had the Linear A and later the Linear B scripts and a numbering system. No other island at that time had writing. Plato says Atlantis was a metalworking state based on copper. Orichalcum, described above, is a copper alloy. Two further passages in the *Critias*:

> ... And they covered with brass, as though with a plaster, all the circumference of the wall which surrounded the outermost circle;

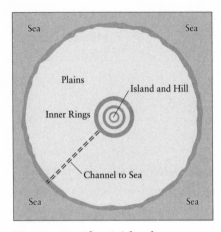

Diagram 1 – Plato's island.

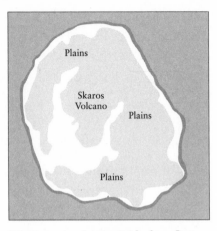

Diagram 4 – Santorini before first eruption c. 15,000 BC.

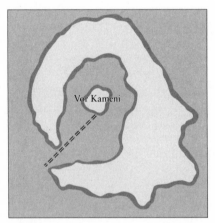

Diagram 5 – 'Minoan' island before 1450 BC eruption showing canal.

Diagram 6 – Island today – after 1450 BC eruption.

and all that of the inner one they covered with tin; and that which encompassing the acropolis itself with orichalcum which sparkled like fire ... [2]

and a little later:

... All the exterior of the temple they covered with silver, save only the pinnacles and these they covered with gold. As to the exterior, they made the roof all of ivory in appearance variegated with gold and silver and orichalcum and all the rest of the walls and pillars and flowers they covered with orichalcum ...

GM: Minoans traded and worked copper, tin, bronze, gold, silver and ivory. Some Minoan buildings had roofs of translucent alabaster to let in the light – as Plato describes it, 'ivory in appearance'.

Only Crete fits the description and the only island people who had the metalworking skills in the period described by Plato were the Minoans.

Plato expands on the metal-making and trading capacity of the Atlantis civilisation:

... For because of their headship they had a large supply of imports from abroad, and the island itself furnished most of the requirements of daily life – metals to begin with, both of the hard kind and the fusible kind which is now known only by name [orichalcum] but was more than a name then, there being mines of it in many places of the island. It brought forth also in abundance all the timbers that a forest provides for the labours of carpenters and of animals it produced a sufficiency, both of tame and wild [elephants] ... [3]

Plato's claim that Atlanteans 'produced a sufficiency, both of tame and wild elephants' seemed to demolish my claim that Atlanteans were Minoans – because obviously, there are no tame elephants to be found on the Mediterranean islands which were part of the Minoan trading empire.

However, I found to my amazement that pygmy elephants were found on Cyprus, Rhodes, the Dodecanese islands, the Cyclades islands, and on Crete, Malta, Sicily and Sardinia in the late Bronze Age (Masseti; Johnson.) their bones have been dated to 2900 BC– 1700 BC. Elephants could not swim to Cyprus. They must have been taken by ship and thus must have been tame.

THE VOLCANIC ERUPTIONS.

Looking towards the sea, but in the centre of the whole island, there was a plain which is said to have been the fairest of all plains and very fertile.

GM: This is the plain on Thera, shown in Diagram 4, before the first major eruption (Diagram 5). When (Plato says 9,000 BC) the island was egg-shaped with a central plain, and almost the same size as Plato describes. The plain was rich in phosphates and nitrates from previous volcanic eruptions and hence very fertile:

Near the plain again, and also in the centre of the island at a distance of about 50 stadia [10,000 yards; 9,000 metres] there was a mountain not very high on any side. In this mountain there dwelt one of the earth-born primeval men of that country whose name was Evenor, and he had a wife named Leucippe, and they had an only daughter who was called Cleito.

GM: This mountain was Skaros volcano, shown on Diagram 4, which was approximately in the centre of the island and 9,000 metres from the east coast. It was not very high compared to the plain – less than 200 metres (220 yards).

The maiden [Cleito] had already reached womanhood, when her father and mother died; Poseidon fell in love with her and had intercourse with her, and breaking the ground enclosed the hill in which she dwelt all around, making alternate zones of sea and land larger and smaller, encircling one another ... so that no man could get to the island for ships and voyages were not as yet ...

GM: Poseidon 'breaking the ground' to seal off the central island, thereby enclosing 'the hill' was in reality the first major volcanic explosion, which would have been passed down in folk memory and would have turned Thera from the island shaped as in Diagram 4 to that shown in Diagram 5. Skaros volcano (4) has become Vor-Kameni island (5). 'For ships and voyages were not as yet.' This signifies that this major explosion was before 6000 BC – when Cretans had ships (they arrived in Crete in 7000 BC).

> He himself, being a god, found no difficulty in making special arrangements for the centre island, bringing up two springs of water from beneath the earth, one of warm water and the other of cold . . .

GM: These are the hot and cold springs in which tourists visiting the central islands of Nea Kameni and Palae Kameni bathe today (Diagram 6). So Plato's description accords with the appearance of Thera both before the central lagoon was flooded and then again, after that point. Plato describes this eruption:

> But at a later time [viz. after Poseidon had surrounded the central island of Nea Kameni with water] there occurred portentous earth-quakes and floods and one grievous day, night befell them . . .

GM: The night is the darkness caused by the explosion of Thera's volcano – the debris would have blotted out the sun, causing crop failures.

> . . . when the whole body of your warriors was swallowed up by the earth and the island of Atlantis was swallowed up by the sea and vanished; wherefore also the ocean at that spot has now become impassable being blocked up by the shoal mud which the island created as it settled down . . .

GM: Plato is here describing the last catastrophic eruption, which changed Thera from what is shown in Diagram 5 to that shown in 6 – more huge chunks of the island have been blown away. A large part of the town now buried beneath modern-day Akrotiri would have gone, as would have some of the settlements on the western part of the main island. Doubtless the shallower parts of the central caldera became a shoal of mud. The sea would have been covered with volcanic tephra, making it appear as mud.

For beginning at the sea, they bored a channel right through the outermost circle which was three plethora in breadth, one hundred feet [30 metres] in depth and 50 stades in length, and thus they made the entrance to it from the sea like that to a harbour by opening up a mouth large enough for the greater ships to sail through.[4]

GM: The route of this channel is shown on Diagrams 5 and 6 from the sea south of Aspro Island, northeast to the central island. Today there is a passage of 30 metres (100 feet) or more in depth and over 6,400 metres (7,000 yards) in length from the sea to Nea Kameni. Plato is probably using some poetic licence; this channel was created by the first major volcanic eruption (compare Diagrams 4 and 5 which let the sea into central Thera) rather than by man – just as the circle of water surrounding the central island was the result of a volcanic eruption, rather than being manmade as Plato claims.

The docks were full of Triremes and naval stores, and all things were quite ready for use.

GM: The busy docks are shown in the Thera fresco as are the triremes and their stores, not least cattle being driven to be loaded on to the ships.

The entire area was densely crowded with habitations; and the canal and the largest of the harbours were full of vessels and merchants coming from all parts who from their numbers, kept up a multitudinous sound of human voices and din and clatter of all sorts, night and day.

GM: Merchants of different nationalities (from their clothes and the colour of their skins) are shown in the frescoes – Libyans, Africans, Minoans, and passengers in the ships with their white gowns.

All of this [dockyard] including the zones and the bridge, they surrounded by a stone wall on every side, placing towers and gates on the bridges where the sea passed in.

GM: The towers and bridges and the surrounding stone wall encircling the dockyard are clearly shown on the Minoan fresco.

> The stone which was used in the work they quarried from underneath the central island, and from underneath the zones, on the outer as well as the inner side. One kind was white, another black and a third red, and as they quarried, they at the same time hollowed out double docks, having roofs carved out of the native rock. Some of these buildings were simple, but in others they put together different stones, varying the colour to please the eye, and to be a natural source of delight.

GM: The white, black and red stones which Plato described appear in the frescoes and are still seen on the cliff faces of Thera today. The coloured stones are also shown in the buildings on the frescoes, as is the subterranean double dock with the roof hollowed out of native rock. This was carved out of the cliff at Red Beach. Just as subterranean docks (for fishing boats) are carved on Santorini today and subterranean houses and restaurants are still being used today.

MYTH, MAGIC AND FINDING AMERICA

Visiting Mycenae on mainland Greece made a big impression on me, not least because it was clear that Heinrich Schliemann had followed Homer to the letter in finding both Troy and Mycenae. Homer had been extraordinarily accurate. This fitted with my own experience, firstly with *1421* and then with *1434*, when it became apparent that the legends of indigenous peoples concerning their ancestors were almost always based on fact. This was especially true in the Americas, where the Indian peoples living on the thousands of miles of Pacific coast that stretch from the Arctic down to South America all maintained that their ancestors came by sea. The same story was later repeated on the North Atlantic coast.

Applying that principle to the history of the Mediterranean, it gradually seemed to me that Plato could have been telling the truth.

I also thought that Greek or Roman historians must have recorded histories of the great Minoan trading empire.

In 1954, at the Naval Training College, Dartmouth, we had been taught Greek and Roman history. So I was in a position to investigate whether Herodotus, Homer or Plato had in fact described Minoan voyages to the Americas. I quickly discovered that authors much more learned than I considered that Homer's *Odyssey* did indeed describe a European Bronze Age fleet circumnavigating the world. I studied some of these histories, notably those of the American ancient history expert Henriette Mertz and several French historians. The problem was that their accounts, which could well be true, were not specific enough – the descriptions could be of America, but they could equally be of the Mediterranean. So reluctantly I discarded them. Then, Marcella found *Atlantis: the Truth Behind the Legend*, by A.G. Galanopoulos and Edward Bacon, the landmark book which opened my eyes.

Over the past forty years, since Professor Galanopoulos and Edward Bacon's book was published, there has been an avalanche of new evidence about the Minoans and their fabulous civilisation. Galanopoulos and Bacon did not have an analysis of the Uluburun wreck and its cargo to go by nor, most important, knowledge of the copper ingots in its hold and their chemical analysis. Nor did they know of the millions of pounds of high-quality copper which disappeared from the mines of Lake Superior in the 3rd millennium BC, apparently into thin air. And they did not have the opportunity to compare the hull of the Uluburun wreck with the Thera frescoes, so as to be in a position to appreciate the magnificent seagoing qualities of Minoan ships. The results of the excavations of Iberian rivers, notably the Bronze Age hoards and Bronze Age ports, were not available to them. The full extent of the Minoan empire, notably the base at Tell el Dab'a in Egypt, was not known to them. As to Stonehenge, the discoveries of Minoan artefacts and skeletons of people who had originally lived in the Mediterranean is very recent. The Minoan pottery in north Germany has only just come to light.

Galanopoulos and Bacon did not know of recent discoveries of Minoan artefacts in copper mines across the world. Neither had they the benefit of the monumental works since published by the Emeritus Professors Sorenson and Johannessen, in which they described enormous levels of transcontinental trade in the Bronze Age. Thanks to new evidence provided by a great many people far more knowledgeable than I, it has been possible to build on Galanopoulos and Bacon's ideas to advance an explanation of Plato's Atlantis which is at once simple and all-embracing.

With the benefits of the enormous amount of recent research we can critically examine Plato's descriptions that concern America:

> ... For the ocean that was at that time navigable; for in front of the mouth which you Greeks call, as you say, 'the Pillars of Heracles' [Straits of Gibraltar] there lay an island which was larger than Libya and Asia together; and it was possible for the travellers of that time to cross from it to the other islands and from the islands to the whole of the continent over against them which encompasses the veritable ocean ... [5]

Here we have Plato describing the Atlantic and America; and stating that the Atlantic is navigable via islands (Canaries and Cape Verde outward bound; Hebrides and Orkneys inward bound).

Plato describes the Minoans' elegant civilisation and says that their empire stretched from across the Atlantic to Europe and embraced the Mediterranean. America lay to the west of the Atlantic '... encompass[ing] the veritable ocean ...', which was navigable in those days.

The huge island in the western Atlantic 'encompassing the ocean' can only refer to America; the fleet returning from America across the Atlantic to attack Athens refers to Minoan ships returning from their voyages with Lake Superior copper.

Such has been the interest in the Atlantis story that a huge amount of investigation and research has been devoted to Thera by professors and experts in all sorts of disciplines – volcanologists, archaeologists, oceanographers, art historians, geographers, meteorologists. It is all fascinating stuff. A selected bibliography of their research is on our website.

THE CONCLUSION THAT'S HARD TO AVOID

The simplest explanation for all these similarities is that Plato was indeed describing the Minoan civilisation. However Plato's story is a conflation of three realities – first, that the Atlantis metropolis was really Santorini; secondly that the island in the Atlantic as big as Libya and Egypt was in fact America; thirdly that Atlantis' manufacturing base and bread basket was Crete. Atlantis was not a single place, but an empire, now lost. Today a visitor taking a colour photograph of the frescoes with him or her can board a boat and sail past Red Beach and in doing so is viewing Plato's Atlantis almost as it was 4,000 years ago.

Truth is indeed stranger than fiction.

Notes to Epilogue

1. Plato, *Critias* 115b
2. Ibid. 116d
3. Ibid. 115e
4. Ibid. 115d
5. Plato, *Timaeus* 25A
6. Masseti and Johnson DL, JSTOR Journal of Biography, vol 7 (1980) p383–398
7. Ibid.

Full details of diagrams on page 338 are contained on ourwebsite. This includes acknowledgements to the German vulcanologists whose plans I have copied.

SELECT BIBLIOGRAPHY

(Full bibliography on website)

Book I: Discovery
The Minoan Civilisation

Alexiou, Stylianos, *Minoan Civilization*, Heraklion, Spyros Alexiou, 1973

Aristotle, *Politics*, 'The olive harvest', 1259a, Cambridge University Press, 1988

Boardman, J., Kenyon, K.M., Moynahan, E.J. and Evans, J.D., 'The Olive in the Mediterranean: Its Culture and Use (and Discussion)', in *Philosophical Transactions of the Royal Society of London, Series B, Biological Sciences*, vol. 275, no. 936, 'The Early History of Agriculture' (27 July 1976), pp. 187–196

Castleden, R., *Minoans: Life in Bronze Age Crete*, London, Routledge, 1992

Crete–Egypt, 3 Millennia of Cultural Interactions, Heraklion Archaeological Museum, 1999

Graves, Robert, *The Greek Myths*, QPD, 1991

Greece, Insight Guides, 1987 and 2008, APA Publications Pte Ltd

Greek Islands, The, Dorling Kindersley, Eye Witness Travel Guide, 1998 and 2007

Hadzi-Vallianou, Despina, *Phaistos*, Athens, Archaeological Receipts Fund, 1989

'Heraklion Archaeological Museum, temporary exhibition', Ministry of Culture, 2007

Knapp, A. Bernard, 'Thalassocracies in Bronze Age Eastern Mediterranean Trade: Making and Breaking a Myth', in *World Archaeology* (Ancient Trade: New Perspectives), Taylor & Francis, vol. 24, no. 3, pp. 332–47, 1993

Michailidou, A., *Knossos – A Complete Guide to the Palace of Minos*, Athens, Ekdotike Athenon S.A., 2006

Museum of Ancient Agora of Athens, Athens, Hellenic Ministry of Culture and Tourism, 1990

National Archaeological Museum – Athens, Athens, Archaeological Receipts Fund, 2009

Sakellarakis, J.A., *Heraklion Museum: Illustrated Guide*, Athens, Ekdotike Athenon S.A., 2006

Tsikritsis, M. (see separate section at the end of the bibliography)

University of Toronto, 'Kommos Excavation Crete' www.fineart.utoronto.ca/kommos/kommosIntroduction.html

Wave that Destroyed Atlantis, The, BBC, www.news.bbc.co.uk/1/hi/6568053.stm

Maps and Charts

Plans in the Southern Kikladhes, Admiralty Chart 1541
Santorini, 1–35,000, Road Editions
Southern Cyclades, 1–190,000, Imray, 2008

Book II: Exploration
Voyages to the Near East

Aruz, Joan, excerpts by, in *Beyond Babylon: Art, Trade and Diplomacy in the Second Millennium* BC, The Metropolitan Museum of Art, New York. Copyright © 2008. Reprinted by permission.

Bard, K., '2006–7 Excavations at Mersa/Wadi Gawasis', paper presented at the 58th annual meeting of the American Research Center in Egypt

Bass, G.F., 'A Bronze Age Shipwreck at Uluburun', in *American Journal of Archaeology*, 1986

Bass, G.F., 'Cape Gelidonya and Bronze Age Maritime Trade', in H.A. Hoffner (ed.), *Orient and Occident*, Alter Orient und Altes Testament, Neukirchener Verlag, 1973

Bass, G.F., 'Cargo From The Age of Bronze: Cape Gelidonya Turkey', in *Beneath the Seven Seas*, London, Thames & Hudson, 2005

Bass, G.F., 'The Construction of a Seagoing Vessel of the late Bronze Age', in H.F. Tzalas (ed.), *Proceedings of the 1st International Symposium on Ship Construction in Antiquity*, Piraeus, 1989

Bass, G.F., 'Evidence of Trade from Bronze Age Shipwrecks', in N.H. Gale (ed.), *Bronze Age Trade in the Mediterranean*, Göteborg, Paul Åströms Förlag, 1991

Bietak, M., 'Avaris and Piramesse: archaeological exploration in the eastern Nile Delta', in *Proceedings of the British Academy 65*, 1979

Bietak, M., 'Some News about Trade and Trade Warfare in Egypt and the Ancient Near East', in *Marhaba*, 3/83

Bietak, M. and Marinatos, N., 'Avaris and the Minoan World', in A. Karetsou (ed.), *Krete-Aigyptos: Politismikoi desmoi trion chilietion*, Athens, 2000

Casson, L., 'Bronze Age Ships. The Evidence of The Theran Wall Paintings', *IJNA*, 4, 1975

Crete–Egypt: Three Thousand Years of Cultural Links, exhibition catalogue, Heraklion and Cairo: Hellenic Ministry of Culture, 2000

Dienekes' Anthropology Blog, 'Minoans in Germany'
http://dienekes.blogspot.com/2008/08/minoans-in-germany.html

Doumas, C. (ed.), *The Wall Paintings of Thera*, Athens, The Thera Foundation, 1992

Duerr, H.P., interviewed in *GEO* magazine
http://www.geo.de/GEO/kultur/geschichte/4669.html

Duerr, H.P., *Rungholt: Die Suche nach einer versunkenen Stadt*, Frankfurt, Insel Verlag, 2005

Gale, N.H., 'Bronze Age Trade in the Mediterranean', in *Studies in Mediterranean Archaeology*, vol. 90, 1991

Gale, N.H., Stos-Gale, Zofia and Maliotis, G., 'Copper Ox-hide Ingots and the Mediterranean Metals Trade', Congress of Cyprus Studies, 2000

Gere, Cathy, *The Tomb of Agamemnon*, London, Profile, 2006

Gimbutas, M., 'East Baltic Amber in the Fourth and Third Millennia', in *Journal of Baltic Studies*, 16, 1985

Gray, D., 'Seewesen', Archeologia Homerica – Lieferungsausgahe, Göttingen, 1974

Grimaldi, D., 'Pushing Back Amber Production', in *Science*, vol. 326, no. 5949, p. 51

Haldane, C., 'Direct Evidence for Organic Cargoes in the Late Bronze Age', in *World Archaeology*, Taylor & Francis, vol. 24, 1993

Hauptmann, Andreas, Maddin, Robert and Prange, Michael, 'On the structure and composition of copper and tin ingots excavated from the shipwreck of Uluburun', in *Bulletin of the American Schools of Oriental Research*, no. 328 (November 2002), pp. 1–30

Hood, Sinclair, *The Minoans; The Story of Bronze Age Crete*, New York, Praeger, 1971

'Mersa/Wadi Gawasis: A Pharaonic Harbor on the Red Sea', exhibition at the Egyptian Museum, Cairo, 2011

Muhly, J.D., 'Sources of Tin and the Beginnings of Bronze Metallurgy', in *American Journal of Archaeology*, vol. 89

Muhly, J.D., Maddin, R., and Stech, T., 'Cyprus, Crete and Sardinia Copper Oxhide Ingots and the Bronze Age Metals Trade', in *Report of the Department of Antiquities*, Cyprus, 1988

Nomikos, P.M., *Sea Voyages: The Fleet Fresco from Thera, and the Punt Reliefs from Egypt*, Piraeus, The Thera Foundation, 2000

Pulak, C., 'Discovering a Royal Ship From The Age of King Tut: Uluburun Turkey', in *Beneath the Seven Seas*, London, Thames & Hudson, 2005

Pulak, C., 'Evidence for Long-distance Trade from the Late Bronze Age Shipwreck at Uluburun', Johannes Gutenberg University, Mainz, 2001

Pulak, C., 'A Hull Construction of the Late Bronze Age Shipwreck at Uluburun', in *INA Quarterly*, 2000

Pulak, C., 'Paired Mortise and Tenon Joints of Bronze Age Seagoing Hulls', in *Boats Ships and Shipyards: Proceedings of the Ninth International Symposium on Ship Construction in Antiquity*, Oxford, Oxbow, 2003

Raban, A., 'The Thera Ships: Another Interpretation', in *American Journal of Archaeology*, vol. 88, 1984

Severin, Tim, *The Jason Voyage*, London, Hutchinson, 1985

Tod, J.M., 'Baltic Amber in the Ancient Near East', in *Journal of Baltic Studies*, 16, 1985

Vergano, Dan, 'In Egyptian desert, a surprising nautical find', *USA Today* www.usatoday.com/tech/science/columnist/vergano/2006–03–05-egyptian-ship – x.htm

Wiener, M.H., 'The Isles of Crete? The Minoan Thalassocracy revisited', in D.A. Hardy, C.G. Doumas, J.A. Sakellarakis and P.M. Warren (eds), *Thera and the Aegean World III*, vol. 1, *Archaeology*, pp. 128–161, London, The Thera Foundation, 1990

Yener, Aslihan K., 'An Early Bronze Age Tin Production Site at Göltepe, Turkey', Oriental Institute, University of Chicago, 1994

India

Abram, D., *The Rough Guide to Kerala*, Rough Guides, London, 2007

Chaudhuri, S.B., *Trade and Civilisation in the Indian Ocean*, Cambridge, Cambridge University Press, 1985

Cherian, P.J. and colleagues, 'Chronology of Pattanam: a multi-cultural port site on the Malabar coast', in *Current Science*, 2009

Frampton, P., *Hidden Kerala: The Travel Guide*, London, MHi Publications, 1997

Frater, A., *Chasing the Monsoon*, London, Penguin, 1991

Hall, R., *Empires of the Monsoon*, London, Harper Collins, 1999

Hindu, The (India's national newspaper), 10 June 2009, 'Prehistoric Cemetery discovered in Kerala'

Kerala Council for Historical Research (KCHR) www.keralahistory.ac.in/

Lothal Guide – http://www.indianetzone.com/6/lothal.htm

Rao, S.R., *Lothal and the Indus Civilisation*, Bombay, Asia Publishing House, 1973

Sreedhara Menon, A., *Kerala History and its Makers*, Madhaven Nayer Foundation, Madras, 1989

Book III: Journeys West

Agricola, G., *De Re Metallica*, New York, Dover, 1950

Baird, W. Sheppard, 'The Early Minoan Colonization of Spain' www.minoanatlantis.com/Minoan–Spain.php

Burl, A., *The Stone Circles of the British Isles*, New Haven, Yale UP, 1976

Cadogan, Gerald, *Palaces of Minoan Crete*, London, Barrie & Jenkins, 1976

Clark, Peter (ed.), *The Dover Bronze Age Boat*, © English Heritage, 2004

Comendador Rey, B., 'Early Bronze Age Technology at Land's End, North Western Iberia', presented at the International Symposium on Science and Technology in Homeric Epics, Olympia, Greece, 2006

Comendador Rey, B., 'The Leiro Hoard (Galicia, Spain): the lonely find?', in *Gold und kult der Bronzezeit*, Germanisches National Museum, Nuremburg, 2003

Comendador Rey, B. and Mèndez, J.L., 'A patina over time: ancient metals conservation in North-Western Iberia', presented at the symposium 'Looking Forward For The Past: Science and Heritage', Tate Modern, London, 2006

Cowen, Richard, University of California, Davis http://mygeologypage.ucdavis.edu/cowen/~gel115/115ch4.html

Cuncliffe, B., *Facing the Ocean: The Atlantic and Its Peoples 8000 BC–AD 1500*, Oxford, Oxford University Press, 2001

Drews, R., *The End of the Bronze Age: Changes in Warfare and the Catastrophe ca. 1200 BC*, Princeton, Princeton University Press, 1995

Goya, Francisco, *La Tauromaquia (1815)*, Prado, Madrid

Henderson, J.C., *The Atlantic Iron Age*, London, Routledge, 2007

Herodotus 3, 115, 'Tin in Europe's extreme west', trans. G. Rawlinson The History of Herodorus, Kessinger Publishing (September 2010)

Hunt Ortiz, M., *Prehistoric Mining and Metallurgy in South West Iberian Peninsula*, Oxford, Archaeopress, 2003

James, D., 'Prehistoric Copper Mining on the Great Ormes Head', *Early Mining in British Isles*, PlasTan-y-Bwich occasional paper No 1, p1–4

Maddin, R., Wheeler, T.S. and Muhly, J.D., 'Tin in the Ancient Near East: Old Questions and New Finds', in *Expedition*, 19.2, 1977

Martin, C., Fernandez-Miranda, M., Fernandez-Posse, D. and Gilman, A., 'The Bronze Age of La Mancha', Universidad Complutense de Madrid, 1993

Muhly, J.D., 'Copper and Tin: the Distribution of Mineral Resources and the Nature of the Metals Trade in the Bronze Age', in *Transactions of the Connecticut Academy of Arts and Sciences*, vol. 43, article 4, 1973

Needham, S., 'The Extent of Foreign Influence on Early Bronze Age axe development in Southern Britain', in Ryan (ed.), *The origins of metallurgy in Atlantic Europe*, Dublin Stationery Office, 1979

Needham, S. et al., *Networks of Contact, Exchange and Meaning: the Beginning of the Channel Bronze Age*, London, British Museum, 2006

O'Brien, W., *Ross Island and the origins of the Irish-British Metallurgy*, publication for 'Ireland in the Bronze Age', Dublin Castle Conference, 1995

Rackham, Oliver, *The Illustrated History of the Countryside*, London, Weidenfeld & Nicolson, 2003

Roberts, E.R.D., *The Great Orme Guide: The Copper Mine, The Prehistoric Period*, 1993

Tylecote, R.F., *The Early History of Metallurgy in Europe*, London, Longman, 1987

Tylecote, R.F., *A History of Metallurgy*, London, The Metals Society, 1976

Tylecote, R.F., *Metallurgy in Archaeology – A Prehistory of Metallurgy in the British Isles*, London, Edward Arnold, 1962

Wertime, T.A. and Muhly, J.D., *Coming of the Age of Iron*, New Haven, Yale UP, 1980

Book IV: Exploring the Heavens
Stonehenge

Atkinson, R.J.C., *Stonehenge*, London, Pelican, 1960

Goskar, Thomas, *British Archaeology*

www.britarch.ac.uk/ba/ba73/feat1.shtml

Hawkins, G., *Beyond Stonehenge*, New York, Harper & Row, 1973

Hawkins, G., *Stonehenge Decoded*, London, Souvenir Press, 1967

Hoyle, F., *On Stonehenge*, San Francisco, Freeman, 1977

Johnson, Anthony, *Solving Stonehenge – The New Key to an Ancient Enigma*, London, Thames & Hudson, 2008

Needham, S. (in course of preparation), *The Archer's Metal Equipment*

Pearson, Parker et al., 'The Age of Stonehenge', in *Antiquity*, vol. 81, 2007

Needham, S., Leese, M.N., Hook, D.R. and Hughes, M.J., 'Developments in the early Bronze Age Metallurgy of southern Britain', in *World Archaeology*, Taylor & Francis, vol. 20, no. 23

Timewatch, BBC, Professors Tim Darvill of the University of Bournemouth and Geoffrey Wainwright, president of the Society of Antiquaries

Book V: The Reaches of Empire
Astronomy and the Calculation of Longitude

Aveni, A., *Stairways to the Stars*, New York, John Wiley, 1997

Brack-Bernsen, Lis, 'Predictions of Phenomena in Babylonian Astronomy', in J.M. Steele & A. Imhausen (eds), *Under One Sky: Astronomy and Mathematics in the Ancient Near East*, Alter Orient und Altes Testament, Muenster, Ugarit-Verlag, 2008

Brack-Bernsen, Lis, 'Some Investigations on the Ephemerides of the Babylonian Moon Texts, System A', in *Centaurus: International Magazine of the History of Science and Medicine*, vol. 24, 1980

Fermor, J. and Steele, J.M., 'The Design of Babylonian Waterclocks', in *International Journal of the History of Mathematics*, vol. 42, issue 3, 2000

Marchant, Jo, *Decoding the Heavens*, Windmill, UK 2009

McKinstry, E. Richard, The Ephemera Society of America, 'So, Just What is Ephemera?' http://www.ephemerasociety.org/news/news-oed.html

Neugebauer, O., *A History of Ancient Mathematical Astronomy*, Berlin, Springer-Verlag, 1975

Neugebauer, O., 'Studies in Ancient Astronomy. VIII. The Water Clock in Babylon', in *Isis*, vol. 37, 1947, reprinted 1983

Oxford English Dictionary (online edition): 'A table showing the predicted (rarely the observed) positions of a heavenly body for every day during a given period.'

Rochberg, F., *The Heavenly Writing*, Cambridge, Cambridge University Press, 2004

Steele, J.M., *A Brief Introduction to Astronomy in the Middle East*, London, SAQI, 2008

Steele, J.M., *Calendars and Years: Astronomy and Time in the Ancient Near East*, Oxford, Oxbow, 2007

Steele, J.M., 'A Commentary on Enuma Anu Enlil 14', in *From the Banks of the Euphrates: Studies in Honor of Alice Louise Slotsky*, Eisenbrauns, Indiana, USA, 2008

Steele, J.M., 'Eclipse Prediction in Mesopotamia', in *Archive For History Of Exact Sciences*, vol. 54, 2000

Steele, J.M., 'Observation, Theory and Practice in Late Babylonian Astronomy: Some Preliminary Observations', in *Astronomy of Ancient Civilizations*, Moscow, Nauka, 2002

Steele, J.M. and Imhausen, A. (eds), *Under One Sky: Astronomy and Mathematics in the Ancient Near East*, Alter Orient und Altes Testament, Muenster, Ugarit-Verlag, 2001

Swerdlow, N.M., 'The Babylonian Theory of the Planets', in *Journal of the American Oriental Society*, vol. 119, 1999. Review by J.M. Steele.

Van der Waerden, B.L., 'Babylonian Astronomy: The Earliest Astronomical Computations', in *Jaarbericht Ex Oriente Lux*, 10, 1948 and *Journal of Near Eastern Studies* 10, 1951

Lunar Eclipses for the 7th Year of Cambyses

(16 July 523 BC and 10 Jan 522 BC)

Steele, J.M., 'Babylonian Predictions of Lunar and Solar Eclipses Times', in *Bulletin of the American Astronomical Society*, vol. 28, 1996

Steele, J.M. and Stephenson, F.R., 'Lunar Eclipse Times Predicted by the Babylonians', in *Journal for the History of Astronomy*, vol. 28, part 2, pp. 119–31

Trans-oceanic Trade

Alfieri, Anastase, 'Les insects de la Tombe de Tutankhamon', in *Bulletin de la Société entomologique d'Egypte*, vol. 24, 1931

Balabanova, S., 'Drugs in Cranial Hair of Pre-Columbian Peruvian Mummies', in *Baessler Archiv NF*, 1992

Balabanova, S., 'First Identification of Drugs in Egyptian Mummies', in *Naturwissenschaften*, 79, 1992

Balabanova S. et al., 'Nicotine and Cotinine in Prehistoric and Recent Bones from Africa and Europe and the Origin of these Alkaloids', in *Homo*, vol. 48, 1997

Carter, G.F., 'Plant Evidence for Early Contacts with America', in *Southwestern Journal of Anthropology*, vol. 6, 1950

Conway, T. and J., *Spirits on Stone – The Agawa Pictographs*, San Luis, Heritage Discoveries, 1990

Coppens, P., 'Copper: A World Trade in 3000 BC?' http://www.philipcoppens.com/copper.html

Dewdney, S. and Kidd, K.E., *Indian Rock Paintings of The Great Lakes*, University of Toronto Press 1962

Drier, Roy W., 'Prehistoric Mining in the Copper Country', in Drier and Du Temple (eds), *Prehistoric Mining in the Lake Superior Region: A Collection of Reference Articles*, privately published

Drier, Roy W. and Du Temple, O., *Prehistoric Mining in the Lake Superior Region: A Collection of Reference Articles*, privately published

Du Temple, O., 'Prehistory's Greatest Mystery: Copper Mines of Ancient Michigan', in *Ancient American*, vol. 5/35

Fonseca, Olympio da, 'Parasitismo e migraçöes humanas pré-históricas: contribuiçöes da parasitologia paro o conhecimento das origens do homem americano', Brazil, University of São Paulo, 1970

Griffin, James B. (ed.), *Lake Superior Copper and the Indians – Miscellaneous Studies of Great Lakes Prehistory*, Ann Arbor, University of Michigan, 1961

Gupta, Shakti M., *Plants in Indian Temple Art*, Delhi, B.R. Publishing, 1996

Heine-Geldern, R.V., 'The Problem of Transpacific Influences in Meso America', in *Handbook of Middle American Indians*, vol. 4, Austin, University of Texas Press

Heine-Geldern, R.V., 'Traces of Indian and Southeast Asiatic Hindu-Buddhist Influences in Mesoamerica', *Proceedings of the 35th International Congress of Americanists*, Mexico, 1962. (Large four-masted Indian ships in prehistoric times were perfectly capable of sailing to the Americas. Wheeled animals show a likely link between India and America from the 3rd millennium BC.)

Heyerdahl, J., *Early Man and the Ocean*, London, George Allen & Unwin, 1978

Hoffman, D., 'Missing: Half-a-billion Pounds of Ancient Copper', in *Ancient American*, vol. 5/35

Jeffreys, M.D.W., 'Pre-Columbian Maize in the Old World: an examination of Portuguese sources', in M.L. Arnott, *Gastronomy: the anthropology of food and food habits*, The Hague, Netherlands, Mouton, 1975, pp 23–66

Johannessen, Carl L., 'Distribution of Pre-Columbian Maize and Modern Maize Names', in Shue Tuck Worg, ed., *Person, Place and Thing: Interpretative and Empincal Essays in Cultural Geography*, volume 31 of *Geoscience and Man*, Geoscience Publications, Louisiana State University, Department of Geography and Anthropology, Baton Rouge, 1992

Johannessen, Carl L., 'Maize Diffused to India Before Columbus came to America', in *Across Before Columbus*, NEARA, 1998

Johannessen, Carl L., 'Pre-Columbian American Sunflower Maize Images in Indian Temples', in *NEARA Journal*, vol. 32, 1998

Johannessen, Carl L. and Parker, Ann Z., 'American Crop Plants in Asia prior to European Contact', in *Proceedings of Conference of Latin Americanist Geographers*, 1989

Leon, Fideas E. et al., 'HLA TransPacific Contacts and Retrovirus', in *Human Immunology*, vol. 42, 1995, p. 349

Meggers, Betty J., 'Yes if by land, no if by sea: the double standard in interpreting cultural similarities', in *American Anthropologist*, 78, 1976

Muhly, J.D., 'An Introduction to Minoan Archaeometallurgy', in *Proceedings of International Symposium*, Crete, 2004

Olsen, Edward J., 'Copper Artefact Analysis with the X-ray Spectrometer', in *American Antiquity*, vol. 28, no. 2, 1962

Riley, Caroll L. et al., *Man Across the Sea: Problems of Pre-Columbian Contacts*, Austin, University of Texas Press, 1971, pp. 219–41

Scherz, J.P., 'Ancient Trade Routes in America's Copper Country', in *Ancient American*, vol. 5/35

Silow, R.A., 'The Problem of Trans-Pacific Migration involved in the origin of the cultivated cottons of the New World', in *Proceedings of Seventh Pacific Science Congress*, vol. 5, New Zealand, 1949

Sorenson, John L., 'The Significance of an Apparent Relationship Between the Ancient Near East and Mesoamerica', in Carroll L. Riley et al.

Sorenson, John L. and Johannessen, Carl L., *World Trade and Biological Exchanges Before 1492*, New York, iUniverse, 2009

Sorenson, John L. and Raish, Martin H., *Pre-Columbian Contact with the Americas across the Oceans: an annotated bibliography*, 2 vols, Utah, Research Press, 1996

Thompson, Gunnar, *American Discovery: Our Multicultural Heritage*, Seattle, Sasquatch, 1999

Thompson, Gunnar, *Secret Voyages to the New World*, Seattle, Misty Isles Press, 2006

Winchell, N.H., 'Ancient Copper Mines of Isle Royale', in *Engineering and Mining Journal*, 32, 1881

Wuthenau, A. von, *Unexpected Faces in Ancient America, 1500 BC–AD 1500: THE HISTORICAL TESTIMONY OF PRE-COLUMBIAN ARTISTS*, New York, Outlet, 1975. (The Babylonian god Humbaba shows up in faces in Veracruz, Chiapas, Columbia and Ecuador. This extraordinary work has been carefully ignored by conventional archaeologists.)

Book VI: The Legacy

Brown, M.D. et al., 'MtDNA Haplogroup X: An Ancient Link Between Europe/Western Asia and North America?', in *American Journal of Human Genetics*, vol. 63, 1998

Cook, R.M., *The Greeks Till Alexander*, London, Thames & Hudson, 1961

Evans, A.J., 'Minoan and Mycenaean Element in Hellenic Life', in *Journal of Hellenic Studies*, 32, 1912

Fitton, J.L., *Minoans*, London, British Museum Press, 2002

Finila, American Journal of human genetics, 2003, November 73 (5), pp 1178–1190, Pub. Med table.
See also Sutton Theory and 'Coming into America: Tracing the Genes', PBS 2004, and 'Stone Age Columbus', BBC 2002, and 'Ice Age Columbus', Discovery 2005, and 'Diffusion of Mf DNA Haplogroup X', American Journal of Human Genetics, 2003.

Galanopoulos, A.G. and Bacon, Edward, *Atlantis: the Truth Behind the Legend*, London, Thomas Nelson, 1969

Graham, J.W., 'The Minoan Unit of Length and Minoan Palace Planning', in *American Journal of Archaeology*, 64, 1960

King, R.J. et al., 'Differential Y-chromosome Anatolian influences on the Greek and Cretan Neolithic', in *Annals of Human Genetics*, vol. 72 (2), 2008, pp. 205–14

Lindsay, Jeff, *Enter Haplogroup X*, www.jefflindsay.com/LDSFAQ/DNA.shtml#x

Morell, Virginia, 'Genes May Link Ancient Eura Sians, Native Americans', Science 24 April 1998: Volume 280, no 5363, p. 520

Reidla, Maere et al., 'Origin and Diffusion of MtDNA Haplogroup X', in *American Journal of Human Genetics*, vol. 73, 2003

Schurr, T.G., 'Mitochondrial DNA and the Peopling of the New World', in *American Scientist*, vol. 18, 2000

Shlush, I. et al., 'The Druze: A Population Genetic Refugium of the Near East', in *Plos One 3*, 2009

Taylour, W., *The Mycenaeans*, London, Thames & Hudson, 1964

Torroni, A. et al., 'Mitochondrial DNA "clock" for the Amerinds and its implications for timing their entry into North America', in *Proceedings of the National Academy of Sciences* (USA), 9, 1994

Triantafyllidis, C. of Thessaloniki's Aristotle University talks about DNA and Minoan genetic origins: http://www.ekathimerini.com

Works of Dr Minas Tsikritsis

(Read after *The Lost Empire of Atlantis* was written)
Calendar Almanac of Cretan-Mycenaean Civilisation, Lawyer's Association of Heraklion, December 2005
Cretan Scripts and the Disc of Phaistos, Secondary Education Office of Heraklion, December 2006
'Egyptian Healing (treatment) Spells in the Language of Keftiu', *Patris* newspaper, 19 April 2005, p. 22
'The Mathematics of the Minoans, Fractions and Decimal System, Geometric Regression', *Eleutherotypia* newspaper, 9 December 2006, p.59
'Medicine in the Bronze Age', *Ichor* magazine, vol. 80, September 2007, pp. 65–6
'Minoans, the Rulers of the Mediterranean', *To Vima* newspaper (science section), 12 August 2007, pp. 27–9
'The Origin of Olympic and Ancient Minoan Games', Prefecture of Heraklion, April 2004

Linear A to the Mountain of Giouktas

'The Disc of Phaistos, A Guide for its Decipherment'
'Linear A – Contributing to the Understanding of an Aegean Script', Vikelea Municipal Library, Heraklion, 2001
'Minoans: The First Cartographers in the World', *Eleutherotypia* newspaper, 24 January 2009
'Plato, Crete, Atlantis and the Holy Mountain of Giouktas', Heraklion, 2008

INDEX